NFPA 72®

National Fire Alarm and Signaling Code

2016 Edition

This edition of NFPA 72, *National Fire Alarm and Signaling Code*, was prepared by the Technical Committees on Fundamentals of Fire Alarm and Signaling Systems, Testing and Maintenance of Fire Alarm and Signaling Systems, Initiating Devices for Fire Alarm and Signaling Systems, Notification Appliances for Fire Alarm and Signaling Systems, Protected Premises Fire Alarm and Signaling Systems, Emergency Communication Systems, Supervising Station Fire Alarm and Signaling Systems, Public Emergency Reporting Systems, and Single- and Multiple-Station Alarms and Household Fire Alarm Systems, released by the Correlating Committee on Signaling Systems for the Protection of Life and Property, and acted on by NFPA at its June Association Technical Meeting held June 22–25, 2015, in Chicago, IL. It was issued by the Standards Council on August 18, 2015, with an effective date of September 7, 2015, and supersedes all previous editions.

This edition of NFPA 72 was approved as an American National Standard on September 7, 2015.

Origin and Development of NFPA 72

The development of NFPA's signaling standards dates back to 1898 with the appointment of the Committee on Thermo-Electric Fire Alarms. The 1905 edition of NBFU 71A, *Rules and Requirements of the National Board of Fire Underwriters for the Construction, Installation, and Use of Signaling Systems Used for the Transmission of Signals Affecting the Fire Hazard as Recommended by the National Fire Protection Association*, and related documents dating back to 1903 were among the first of numerous signaling standards published in conjunction with the National Fire Protection Association. The descendants of these earlier standards have been consolidated into the *National Fire Alarm Code, NFPA 72.*

The first edition of the *National Fire Alarm Code*, published in 1993, was a consolidation of the 1989 edition of NFPA 71, *Standard for the Installation, Maintenance, and Use of Signaling Systems for Central Station Service*; the 1990 edition of NFPA 72, *Standard for the Installation, Maintenance, and Use of Protective Signaling Systems*; the 1990 edition of NFPA 72E, *Standard on Automatic Fire Detectors*; the 1989 edition of NFPA 72G, *Guide for the Installation, Maintenance, and Use of Notification Appliances for Protective Signaling Systems*; the 1988 edition of NFPA 72H, *Guide for Testing Procedures for Local, Auxiliary, Remote Station, and Proprietary Protective Signaling Systems*; and the 1989 edition of NFPA 74, *Standard for the Installation, Maintenance, and Use of Household Fire Warning Equipment.* Many of the requirements of these standards were identical or very similar. The recommendations that were taken from the guides (NFPA 72G and NFPA 72H) were changed to mandatory requirements.

The 1996 edition of *NFPA 72* incorporated many changes of a technical nature. These changes related to issues such as the Americans with Disabilities Act, software testing, fire modeling, and communications.

The 1999 edition represented a major change in code content and organization. The chapters were arranged to facilitate user friendliness and provide a logical structure. A new chapter on public fire reporting was added, and many technical revisions were made. Annex B (formerly Appendix B) was streamlined to facilitate ease of use, many unenforceable terms were removed, and Chapter 3 was reorganized to facilitate a more logical approach.

The 2002 edition reflected an extensive editorial revision of the Code to comply with the latest edition of the *Manual of Style for NFPA Technical Committee Documents.* These revisions included the addition of three administrative chapters at the beginning of the Code: "Administration," "Referenced Publications," and "Definitions." Eight technical chapters followed the administrative chapters in the same sequence as in the 1999 edition. Other editorial revisions included the breakout of paragraphs with multiple requirements into individually numbered paragraphs for each requirement, the minimization of use of exceptions, the use of consistent headings for sections and section subdivisions, and reorganization to limit paragraph numbering to six digits.

The 2002 edition contained a number of technical revisions throughout the Code. These included a major revision of the power supply requirements; a new requirement addressing impairments to fire alarm systems; additional requirements concerning the review and approval of performance-based detection system designs; revision of the rules for system survivability from attack by fire; the introduction of rules for an alternate approach for audible signaling; the addition of requirements to address performance-based designs for visible signaling; the relocation of testing and maintenance requirements for single- and multiple-station alarms and household fire alarm systems to the testing and maintenance chapter; and revisions to re-establish the prescriptive rules for household fire-warning equipment from the 1996 edition of the Code.

The 2007 edition contained a number of technical revisions including many to accommodate new technology and to take advantage of new research. Changes were made to better address the integration of mass notification systems and other systems with fire alarm systems. Revisions were also made in several areas of the Code for clarity and to enhance its usability.

Some of the more significant revisions in the 2007 edition addressed protection of fire alarm control units, personnel qualification, heat detector response time, smoke detector spacing, smoke detection in ducts, detectors that use multiple sensing inputs, video image smoke and flame detection, synchronization of visible notification appliances, exit marking audible notification appliances, tactile notification appliances, different types of protected premises fire alarm system, and in-building enhancement systems for firefighter radio communications.

The 2007 edition also included significant changes to the requirements for smoke alarms in residential applications, revisions to require the interconnection of smoke alarms for existing occupancies, revisions to require additional smoke alarms for larger dwelling units, and revisions to allow voice messages to be included as a part of the smoke alarm notification signal.

Revisions in the 2007 edition to enhance and clarify the Code included those that address suppression system inputs to the fire alarm system, emergency/voice alarm communications systems, fire alarm system interface with elevator systems, and the means to indicate central station service. In addition, a complete revision of the Record of Completion Form along with examples of filled-out forms were provided.

The 2007 edition also included the addition of two new annexes — one to provide guidance for the design of mass notification systems and one to replace previous annex material for the design of fire service interfaces with a separate industry standard.

The 2010 edition of the Code presented a major change in the scope and organization of the document. This was reflected in the new title, *National Fire Alarm and Signaling Code*. The broader scope of the Code included emergency communications systems in addition to the traditional scope of fire alarm systems. A new chapter on emergency communications systems (ECS) was added to provide requirements for a variety of systems used for communication of information in various emergency situations. The ECS chapter included new systems such as in-building mass notification systems, wide-area mass notification systems, distributed recipient mass notifications systems, two-way radio communications enhancement system, and area of refuge emergency communications systems. The ECS chapter also included two systems formerly located in the chapter on protected premises fire alarm systems: (in-building fire) emergency voice/alarm communications systems, and two-way in-building wired (telephone) emergency services communication systems.

Two other new chapters were added in the 2010 edition. The new chapter on circuits and pathways included requirements and information formerly from the chapters on fundamentals of fire alarm systems and from the chapter on protected premises fire alarm systems. This new chapter provided circuit and pathway performance (class) designations and pathway survivability level designations as well as general wiring requirements presented in a format that allowed use by any type of system covered in the Code. The new chapter on emergency control functions and interfaces included requirements and information formerly contained in the chapter of protected premises fire alarm systems. In this chapter the term *fire safety function* generally was replaced with the term *emergency control function* to reflect the potentially broader application beyond just fire alarm systems. This new chapter also included new provisions for first-responder-use elevators and elevators for occupant-controlled evacuation.

The 2010 edition was substantially reorganized to accommodate the new chapters in a logical order with reserved chapter numbers included to minimize the potential for further renumbering in the future. The overall organization included administrative chapters, support chapters, and system chapters as well as numerous annexes for usability.

Significant revisions were also made throughout the 2010 edition to reflect the broader application of the Code to emergency communication systems. These included revisions within the chapter on fundamentals to address power supply requirements, signal priorities, signal distinction and documentation requirements; revisions to the protected premises chapter to better accommodate non–fire alarm systems in combination systems; revisions to the supervising stations chapter and public emergency reporting system chapter to allow their use for emergency communications systems; and revisions to the testing and maintenance chapter to incorporate requirements for the inspection, testing, and maintenance of mass notification systems and two-way radio communications enhancement systems.

In addition to the content of the new chapters, the 2010 edition included significant technical changes. These included new requirements for signaling to the deaf and hard of hearing, new requirements and guidance for the design, installation, and testing of voice communications systems to ensure voice intelligibility, and extensive revision of requirements for the installation of smoke detectors in both level and sloped joist and beam ceiling applications.

Revisions to the supervising station chapter in the 2010 edition included the retirement of four legacy transmission technologies no longer being installed: active multiplex transmission systems, McCulloh systems, directly connected noncoded systems, and private microwave systems. The subsection on "Other Transmission Technologies" was relocated to become the default subsection for supervising stations communications methods.

Modifications in the 2010 edition to the chapter on single- and multiple-station alarms and household fire alarms systems included new provisions to address the interconnection of smoke alarms using wireless technology, new provisions for signaling to those with hearing loss, and new requirements and guidance for the placement of smoke alarms and smoke detectors.

The 2010 edition included two new guidance annexes, Annex C on system performance and design, and Annex D on speech intelligibility.

Prior editions of this document have been translated into languages other than English, including Spanish.

The 2013 edition of the Code was built on the organizational changes made in the 2010 edition. A new Chapter 7, "Documentation," which was added to improve the usability of the document, provided a central location for all the documentation requirements of the Code. In some cases, the documentation provisions were contained directly in the new chapter. In other cases, references were provided to the locations of documentation requirements contained in other chapters. As an example, the new chapter contained the minimum documentation requirements that applied to any system covered by the Code, while additional document requirements that might apply from other parts of the Code or from other governing laws, codes, or standards were listed with an appropriate reference. The Record of Completion and Record of Inspection, Testing, and Maintenance forms were included at the end of the chapter and were completely revised to be easier to use with a basic form for straightforward systems and supplemental forms for more complex systems.

Chapter 10, "Fundamentals," was reorganized for the 2013 edition to provide a more user-friendly flow of requirements. In addition, requirements for circuit monitoring found in the previous edition of Chapter 10 were relocated to Chapter 12, "Circuits and Pathways," a more logical location.

Extensive usability changes also were made in the inspection and testing tables of Chapter 14, "Inspection, Testing, and Maintenance." The visual inspection table was updated, adding new inspection methods for each component along with the inspection frequency. The test methods and test frequency tables were combined into a single table so that the test method appeared along with the test frequency for each component. The component listings in both tables were reorganized and coordinated so that components and equipment were easier to find.

The 2013 edition of the Code also included many technical updates. Among those updates were changes in Chapter 10, "Fundamentals," requiring supervising station operators and fire alarm system service providers to report to the authority having jurisdiction certain conditions of system impairment. Requirements for inspection, testing, and service personnel qualifications were updated to better reflect the level of qualification needed for each type of activity. Changes were also made in Chapter 18, "Notification Appliances," requiring documentation of the locations that require audible notification appliances as well as documentation of the audibility levels that must be produced. Area of coverage requirements were added for visible notification appliances. Changes were made in Chapter 21, "Emergency Control Function Interfaces," to address requirements for elevator recall when sprinklers are installed in elevator pits. The requirements for occupant evacuation elevators were completely revised to coordinate with changes being made in ASME A.17.1/B44, *Safety Code for Elevators and Escalators.* Changes were made in Chapter 24, "Emergency Communications Systems," to address the use of microphones, to address the use of textual and graphical visible notification appliances for primary or supplemental notification, and to update the requirements for emergency command centers. Changes were made in Chapter 26, "Supervising Station Alarm Systems," to address alarm signal verification, alarm signal content, and restoration of signals. Those changes were made in part to help emergency responders better manage issues related to unwanted alarms. In addition, new definitions for unwanted alarms were added to more precisely identify the sources of those alarms. Changes were also made to update the communications methods addressed in Chapter 26. Among them were changes to the supervision interval requirements for communications paths and changes to the types of transmission means that can be used for the second channel of a DACT. Changes were made in Chapter 29, "Single- and Multiple-Station Alarms and Household Fire Alarm Systems," to address the connection of sprinkler waterflow switches to multiple-station alarms and to add new requirements addressing the smoke alarm resistance to common nuisance sources.

The 2016 edition makes many changes relative to documentation. Chapter 7 revises and adds items to the minimum documentation, documentation for new emergency communications systems, and software documentation requirements and addresses review of electronic documentation media formats. Requirements for documentation of qualifications for the system designer and personnel who program systems are more clearly addressed while providing for the allowance for system design trainees. New criteria have been added for plans examiners and inspectors.

Perhaps the most significant changes to the Code pertain to wiring. The 2016 edition adds Class N, which addresses internet infrastructures for alarm and signaling systems; pathway performance and installation criteria are provided. Class A and Class X pathway separation requirements have been revised to specifically address emergency control function interface devices controlled by the fire alarm system on those circuits. Level 2 and Level 3 pathway survivability requirements have been revised, providing flexibility of use and to address other "fire-resistive" methods.

The 2016 edition of the Code adds language relative to recalled equipment observed during inspection and testing and clarifies the intent of periodic visual inspections relative to building or other changes that could affect the performance of the system.

With the exception of reference and requirements pertaining to survivability, requirements for the design, installation, testing, and maintenance of in-building emergency radio communications enhancement systems have been relocated to NFPA 1221, *Standard for the Installation, Maintenance, and Use of Emergency Services Communications Systems.*

Chapter 17 revises the requirements for "total coverage" and expands its annex language to address general consideration for elevator shafts and enclosed stairways. The requirements for placement of smoke detectors used for door release also have been revised to provide additional flexibility in locating detectors.

Chapter 24 has been restructured, proving greater user friendliness while expanding the section on risk analysis. Emphasis is placed on the importance of effective message development. Annex G, "Guidelines for Emergency Communications Strategies for Buildings and Campuses," based on the National Institute of Standards and Technology and Fire Protection Research Foundation research has been added to the 2016 edition.

Language has been changed in Chapter 26 and now requires that when multiple communication paths are used for performance-based technologies or the two transmission means for a digital alarm communicator transmitter that they be arranged to avoid a single point of failure.

Chapter 29 now has requirements pertaining to remote resetting and silencing of a fire alarm control unit from other than the protected premises for a minimum of 4 minutes from the initial activation of the fire alarm signal. Smartphones and Internet access to almost any device have made remote access to residential equipment possible. The ability to establish remote access to a fire alarm system is addressed. Also, a new requirement establishes that, where a communication or transmission means other than DACT is used, all equipment necessary to transmit an alarm signal must be provided with a minimum of 24 hours of secondary power capacity.

Correlating Committee on Signaling Systems for the Protection of Life and Property

Robert P. Schifiliti, *Chair*
R. P. Schifiliti Associates, Inc., MA [SE]

Lee F. Richardson, *Administrative Secretary*
National Fire Protection Association, MA

Douglas M. Aiken, Lakes Region Mutual Fire Aid, NH [U]
Rep. International Municipal Signal Association
Andrew G. Berezowski, Honeywell Inc., CT [M]
Rep. National Electrical Manufacturers Association
Art Black, Carmel Fire Protection Associates, CA [E]
David L. Boswell, Hughes Associates/RJA Group, CO [SE]
John C. Fannin III, SafePlace Corporation, DE [U]
Louis T. Fiore, L. T. Fiore, Inc., NJ [IM]
Rep. Central Station Alarm Association
Bruce Fraser, Fraser Fire Protection Services, MA [SE]
Vic Humm, Vic Humm & Associates, TN [SE]
Peter A. Larrimer, U.S. Department of Veterans Affairs, PA [U]
A. M. Fred Leber, LRI Fire Protection Engineering, Inc., Canada [SE]
James M. Mundy, Jr., Asset Protection Associates, Ltd., NY [M]
Rep. Automatic Fire Alarm Association, Inc.
Lynn Nielson, City of Henderson, NV [E]
Thomas F. Norton, Norel Service Company, Inc., MA [IM]
Rep. U.S. Naval Historical Center
(VL to Document: 72)
Thomas J. Parrish, Telgian Corporation, MI [U]
Rodger Reiswig, Tyco/SimplexGrinnell, FL [M]
Lawrence J. Shudak, UL LLC, IL [RT]
Tom G. Smith, Oklahoma City, OK [IM]
Rep. National Electrical Contractors Association
Earl P. Valois, New Orleans Fire Department, LA [E]
Rep. International Association of Fire Chiefs
Jeffery G. Van Keuren, UTC Climate, Controls & Security/Edwards, FL [M]

Alternates

Leonard Belliveau, Jr., Hughes Associates/RJA Group, RI [SE]
(Alt. to D. L. Boswell)
Anthony J. Capowski, Tyco/SimplexGrinnell, MA [M]
(Alt. to R. Reiswig)
Louis Chavez, UL LLC, IL [RT]
(Alt. to L. J. Shudak)
Raymond A. Grill, Arup, DC [SE]
(Alt. to A. M. Fred Leber)
Thomas P. Hammerberg, Automatic Fire Alarm Association, Inc., GA [M]
(Alt. to J. M. Mundy, Jr.)
James Ludden, Bolivar City Fire Department, MO [E]
(Alt. to E. P. Valois)
Jack McNamara, Bosch Security Systems, NY [M]
(Alt. to A. G. Berezowski)
Frank L. Van Overmeiren, FP&C Consultants, Inc., IN [SE]
(Alt. to V. Humm)

Nonvoting

Merton W. Bunker, Jr., U.S. Department of State, VA [U]
Rep. TC on Protected Premises Fire Alarm Systems
Laurence J. Dallaire, U.S. Architect of the Capitol, DC [E]
Rep. TC on Household Fire Alarm Systems
Manuelita E. David, Aon Fire Protection Engineering, CA [I]
Rep. TC on Fundamentals of Fire Alarm Systems
Jeffrey G. Knight, City of Newton Fire Department, MA [U]
Rep. TC on Public Fire Reporting Systems
David O. Lowrey, City of Boulder Fire Rescue, CO [E]
Rep. TC on Notification Appliances for Fire Alarm Systems
J. Jeffrey Moore, Hughes Associates/RJA Group, OH [SE]
Rep. TC on Testing & Maintenance of Fire Alarm Systems
Wayne D. Moore, Hughes Associates/RJA Group, RI [SE]
Rep. TC on Emergency Communication Systems
Daniel J. O'Connor, Aon Fire Protection Engineering, IL [I]
Rep. TC on Initiating Devices for Fire Alarm Systems
Warren E. Olsen, Fire Safety Consultants, Inc., IL [E]
Rep. TC on Supervising Station Fire Alarm Systems
Martin H. Reiss, Hughes Associates/RJA Group, MA [SE]
Rep. Safety to Life Correlating Committee
Dean K. Wilson, Hughes Associates/RJA Group, PA [SE]
(Member Emeritus)

Lee F. Richardson, NFPA Staff Liaison

This list represents the membership at the time the Committee was balloted on the final text of this edition. Since that time, changes in the membership may have occurred. A key to classifications is found at the back of the document.

NOTE: Membership on a committee shall not in and of itself constitute an endorsement of the Association or any document developed by the committee on which the member serves.

Committee Scope: This Committee shall have primary responsibility for documents on the installation, performance, maintenance, testing, and use of signaling components and signaling systems for the protection of life, property and mission continuity.

Technical Committee on Emergency Communication Systems

Wayne D. Moore, *Chair*
Hughes Associates/RJA Group, RI [SE]

Daniel P. Finnegan, *Secretary*
Siemens Industry, Inc., IL [M]
Rep. National Electrical Manufacturers Association

Oded Aron, Port Authority of New York & New Jersey, NJ [U]
Peter Binkley, Evax Systems, Inc., CT [M]
Daniel Bridgett, U.S. Department of the Navy, CA [E]
Thomas M. Chambers, Vector Security Inc., PA [IM]
Rep. Central Station Alarm Association
Joe L. Collins, Dallas/Fort Worth International Airport, TX [U]
Joseph Dafin, U.S. General Services Administration, DC [U]
John C. Fannin III, SafePlace Corporation, DE [U]
Bruce Fraser, Fraser Fire Protection Services, MA [SE]
John S. Fuoto, AMEC Environment & Infrastructure, SC [SE]
Charles E. Hahl, The Protection Engineering Group, Inc., VA [SE]
Raymond N. Hansen, U.S. Department of the Air Force, FL [U]
Waymon Jackson, University of Texas at Austin, TX [U]
Scott Lacey, Lacey Fire Protection Engineering, AR [SE]
Minfei M. Leng, Bird Technologies Group, NY [M]
Derek D. Mathews, UL LLC, IL [RT]
James Mongeau, Space Age Electronics, Inc., MA [M]
Rep. Automatic Fire Alarm Association, Inc.
Michael Pallett, Telecor Inc., Canada [M]
Thomas J. Parrish, Telgian Corporation, MI [IM]
Joseph Ranaudo, AFA Protective Systems, Inc., NY [IM]
Rodger Reiswig, Tyco/SimplexGrinnell, FL [M]
Jeffrey A. Scott, FP&C Consultants, Inc., MO [SE]
Aviv Siegel, AtHoc, Inc., CA [M]
James P. Simpson, Electrical Training Alliance, MN [L]
Rep. International Brotherhood of Electrical Workers
Robert Anthony Ulizio, Georgetown University, DC [U]
Andrew B. Woodward, Arup Fire, MA [SE]

Alternates

Laura E. Doyle, U.S. General Services Administration, DC [U]
(Alt. to J. Dafin)
Jon M. Evenson, Hughes Associates/RJA Group, IL [SE]
(Alt. to W. D. Moore)
Aleksandr Frenkel, Port Authority of New York & New Jersey, NJ [U]
(Alt. to O. Aron)
Raymond A. Grill, Arup, DC [SE]
(Alt. to A. B. Woodward)
Paul E. Macknis, U.S. Department of the Navy, CA [E]
(Alt. to D. Bridgett)
David Madole, University of Texas at Austin, TX [U]
(Alt. to W. Jackson)
Denise L. Pappas, Valcom, Inc., VA [M]
(Alt. to D. P. Finnegan)
Yatin J. Patel, UL LLC, IL [RT]
(Alt. to D. D. Mathews)
Sean C. Remke, 1st Fire Protection & Code Consultants, LLC, IN [SE]
(Alt. to J. A. Scott)
Joseph Ripp, Interface Engineering, OR [M]
(Alt. to J. Mongeau)
Christo Spyridis, Telecor Inc., Canada [M]
(Alt. to M. Pallett)
Mark J. Sylvester, Telgian Corporation, AZ [IM]
(Alt. to T. J. Parrish)
Andrew P. Valente, AMEC Environment & Infrastructure, FL [SE]
(Alt. to J. S. Fuoto)
Larry D. Watson, American Professional Services, Inc., OK [IM]
(Alt. to T. M. Chambers)

Lee F. Richardson, NFPA Staff Liaison

This list represents the membership at the time the Committee was balloted on the final text of this edition. Since that time, changes in the membership may have occurred. A key to classifications is found at the back of the document.

NOTE: Membership on a committee shall not in and of itself constitute an endorsement of the Association or any document developed by the committee on which the member serves.

Committee Scope: This committee shall have primary responsibility for documents on the risk analysis, design, application, installation, and performance of emergency communications systems and their components. Public emergency services communications systems covered by NFPA 1221 are outside the scope of this committee except where they interface with in-building bi-directional amplifiers and where trouble and supervisory signals are intended to be monitored by the building fire alarm system.

Technical Committee on Fundamentals of Fire Alarm and Signaling Systems

Manuelita E. David, *Chair*
Aon Fire Protection Engineering, CA [I]

Jeffrey S. Hancock, *Secretary*
Valero, TX [U]

Mark J. Aaby, Koffel Associates, Inc., MD [SE]
James C. Abner, Artesia Fire Department, NM [E]
Andrew G. Berezowski, Honeywell Inc., CT [M]
Rep. National Electrical Manufacturers Association
Robert A. Bonifas, Alarm Detection Systems, Inc., IL [IM]
Rep. Central Station Alarm Association
Shane M. Clary, Bay Alarm Company, CA [IM]
Daniel G. Decker, Safety Systems, Inc., MI [IM]
James DiTaranto, Commercial Electrical Systems, FL [IM]
Sanford E. Egesdal, Egesdal Associates PLC, MN [SE]
Tommy L. Farr, JATC of Southern Nevada, NV [L]
Rep. International Brotherhood of Electrical Workers
David W. Frable, U.S. General Services Administration, IL [U]
Daniel J. Gauvin, Tyco Fire Suppression & Building Products, MA [M]
David Goodyear, Seneca College, Canada [SE]
Kevin M. Green, Pyro-Comm Systems, Inc., CA [IM]
Kimberly A. Gruner, Fike Corporation, MO [M]
Rep. Fire Suppression Systems Association
Scott Jacobs, ISC Electronic Systems, Inc., CA [IM]
Jon Kapis, Hughes Associates/RJA Group, CA [SE]
Walter J. Kessler, Jr., FM Approvals, MA [I]
A. M. Fred Leber, Leber/Rubes Inc. (LRI), Canada [SE]
Chester S. Maciaszek, Savannah River Nuclear Solutions, LLC, SC [U]
Richard A. Malady, Fire Fighter Sales & Service Company, PA [IM]
Rep. National Association of Fire Equipment Distributors
Jack McNamara, Bosch Security Systems, NY [M]
James M. Mundy, Jr., Asset Protection Associates, Ltd., NY [M]
Rep. Automatic Fire Alarm Association, Inc.
Louis Nash, U.S. Coast Guard, DC [E]
Thomas F. Norton, Norel Service Company, Inc., MA [IM]
David J. Stone, UL LLC, IL [RT]
Emily Troyanski, Intertek Testing Services, NJ [RT]
Todd W. Warner, Brooks Equipment Company, Inc., NC [M]
Rep. Fire Equipment Manufacturers' Association

Alternates

Jason W. Butler, Savannah River Nuclear Solutions, LLC, GA [U]
(Alt. to C. S. Maciaszek)
Bob Elliott, FM Approvals, MA [I]
(Alt. to W. J. Kessler, Jr.)
Erinc Eslik, Intertek Testing Services, CT [RT]
(Alt. to E. Troyanski)
John Houlihan, Encore Fire Protection, RI [M]
(Alt. to J. M. Mundy, Jr.)
Daniel M. Kester, Battelle, WA [U]
(Alt. to Je. S. Hancock)
William E. Koffel, Koffel Associates, Inc., MD [SE]
(Alt. to M. J. Aaby)
Maria B. Marks, Siemens Industry, Inc., MD [M]
(Alt. to A. G. Berezowski)
John McCamish, NECA IBEW Electrical Training Center, OR [L]
(Alt. to T. L. Farr)
John H. Miller, U.S. Coast Guard, DC [E]
(Alt. to L. Nash)
Rich Neves, Pyro-Comm Systems, Inc., CA [IM]
(Alt. to K. M. Green)
Timothy R. Oates, Tyco/SimplexGrinnell, DE [M]
(Alt. to D. J. Gauvin)
Allan Sanedrin, UL LLC, IL [RT]
(Alt. to D. J. Stone)
Richard A. Savicki, Safety Systems, Inc., MI [IM]
(Alt. to D. G. Decker)
David M. Secoda, Aon Fire Protection Engineering, CA [I]
(Alt. to M. E. David)
Ed Vaillancourt, E & M International, Inc., NM [M]
(Alt. to K. A. Gruner)
William F. Wayman, Jr., Hughes Associates/RJA Group, MD [SE]
(Alt. to J. Kapis)
Robert A. Williams, II, Vector Security Inc., VA [IM]
(Alt. to R. A. Bonifas)

Lee F. Richardson, NFPA Staff Liaison

This list represents the membership at the time the Committee was balloted on the final text of this edition. Since that time, changes in the membership may have occurred. A key to classifications is found at the back of the document.

NOTE: Membership on a committee shall not in and of itself constitute an endorsement of the Association or any document developed by the committee on which the member serves.

Committee Scope: This Committee shall have primary responsibility for documents on common system fundamentals for fire alarm and signaling systems, requirements for approvals, power supplies, equipment performance, system documentation, and compatibility.

Technical Committee on Single- and Multiple-Station Alarms and Household Fire Alarm Systems

Laurence J. Dallaire, *Chair*
U.S. Architect of the Capitol, DC [E]

Edward M. Fraczkowski, *Secretary*
EBL Engineers, LLC, MD [SE]

H. Wayne Boyd, U.S. Safety & Engineering Corporation, CA [IM]
Rep. California Automatic Fire Alarm Association Inc.
David E. Christian, Gentex Corporation, MI [M]
Rep. Automatic Fire Alarm Association, Inc.
Thomas G. Cleary, National Institute of Standards & Technology, MD [RT]
Timothy K. Dedear, City of Farmers Branch Fire Department, TX [E]
Wendy B. Gifford, Consultant, IL [SE]
Daniel T. Gottuk, Hughes Associates/RJA Group, MD [SE]
Richard Hoffman, Intertek, IL [RT]
Mark A. Kittle, Town of Snowmass Village, CO [E]
Anna Kryagin, Port Authority of New York & New Jersey, NJ [U]
Thomas J. McNelis, Jarden Safety & Security/BRK Brands, IL [M]
Rep. National Electrical Manufacturers Association
David E. Mills, UL LLC, IL [RT]
Kim R. Mniszewski, FX Engineering, Inc., IL [SE]
Cory Ogle, Code Consultants, Inc., MO [SE]
Jeffrey L. Okun, Nuko Security, Inc., LA [IM]
Stephen M. Olenick, Combustion Science & Engineering, Inc., MD [SE]
Steven Orlowski, National Association of Home Builders, DC [U]
Forrest J. Pecht, U.S. Access Board, DC [C]
Timothy A. Rader, ADT Security Services, Inc., FL [M]
Larry Ratzlaff, UTC/Kidde Safety, IL [M]
Richard M. Simpson, Vector Security Inc., PA [IM]
Rep. Central Station Alarm Association
Thomas R. Waldron, Aon Fire Protection Engineering, CA [I]

Alternates

Oded Aron, Port Authority of New York & New Jersey, NJ [U]
(Alt. to A. Kryagin)
Andrew Blum, Exponent, Inc., GA [SE]
Daniel Buuck, National Association of Home Builders, DC [U]
(Alt. to S. Orlowski)
Manuelita E. David, Aon Fire Protection Engineering, CA [I]
(Alt. to T. R. Waldron)
Joshua B. Dinaburg, Hughes Associates/RJA Group, MD [SE]
(Alt. to D. T. Gottuk)
Rick Heffernan, SDi, NJ [M]
(Alt. to T. J. McNelis)
Nasir Hussain, Combustion Science & Engineering, Inc., MD [SE]
(Alt. to S. M. Olenick)
Paul T. Kahle, Code Consultants, Inc., MO [SE]
(Alt. to C. Ogle)
Jeffery P. McBride, EBL Engineers, LLC, MD [SE]
(Alt. to E. M. Fraczkowski)
Ken Mott, UTC/Kidde Safety, CO [M]
(Alt. to L. Ratzlaff)
John L. Parssinen, UL LLC, IL [RT]
(Alt. to D. E. Mills)
Richard Jay Roberts, Honeywell Life Safety, IL [M]
(Alt. to D. E. Christian)
Raymond Smolenski, ADT Security Services, Inc., FL [M]
(Alt. to T. A. Rader)
Samuel T. (Ted) Stoler, Vector Security Inc., PA [IM]
(Alt. to R. M. Simpson)
Emily Troyanski, Intertek Testing Services, NJ [RT]
(Alt. to R. Hoffman)

Nonvoting

Arthur S. Lee, U.S. Consumer Product Safety Commission, MD [C]
Maurice M. Pilette, Mechanical Designs Ltd., MA [SE]
Rep. TC on Residential Sprinkler Systems

Lee F. Richardson, NFPA Staff Liaison

This list represents the membership at the time the Committee was balloted on the final text of this edition. Since that time, changes in the membership may have occurred. A key to classifications is found at the back of the document.

NOTE: Membership on a committee shall not in and of itself constitute an endorsement of the Association or any document developed by the committee on which the member serves.

Committee Scope: This Committee shall have primary responsibility for documents on the performance, installation, operation, inspection, testing, maintenance, and use of single- and multiple-station alarms and household alarm systems for fire warning.

Technical Committee on Initiating Devices for Fire Alarm and Signaling Systems

Daniel J. O'Connor, *Chair*
Aon Fire Protection Engineering, IL [I]

Wayne J. Aho, Xtralis, Inc., MA [M]
Mark S. Boone, Dominion Resources Services Inc., VA [U]
Rep. Edison Electric Institute
Jan Braam, UTC Climate, Controls & Security/Edwards, FL [M]
Rep. National Electrical Manufacturers Association
John A. Chetelat, Honeywell Life Safety Group, CT [M]
Rep. Fire Suppression Systems Association
John M. Cholin, J. M. Cholin Consultants Inc., NJ [SE]
Kenneth W. Dungan, Performance Design Technologies, TN [SE]
Gary P. Fields, The Protectowire Company, Inc., MA [M]
Robert A. Hall, R. A. Hall & Associates, NJ [SE]
Loren L. Leimer, Hochiki America Corporation, CA [M]
Rep. Automatic Fire Alarm Association, Inc.
Michael Lessar, Reading Fire Department, PA [E]
Norbert W. Makowka, National Association of Fire Equipment Distributors, IL [IM]
Rep. National Association of Fire Equipment Distributors
Chris Marrion, Marrion Fire & Risk Consulting PE, LLC, NY [SE]
Justin D. Merrick, S&S Sprinkler Company, LLC, AL [IM]
Samuel M. Miller, BP Exploration (Alaska) Inc., AK [U]
Ovid E. Morphew, Jr., Design/Systems Group, TX [IM]
Rep. National Independent Fire Alarm Distributors Assn.
James W. Mottorn, II, Bosch Security Systems, NY [M]
Lynn Nielson, City of Henderson, NV [E]
Ronald D. Ouimette, Siemens Building Technologies, Inc., NJ [M]
John L. Parssinen, UL LLC, IL [RT]
Martin H. Reiss, Hughes Associates/RJA Group, MA [SE]
David L. Royse, Potter Electric Signal Company, MO [M]
James R. Schario, Electrical Industry Training Center (IBEW/NECA), MO [L]
Rep. International Brotherhood of Electrical Workers
Kenneth R. Schneider, Sprinkler Fitters LU 268 JATC, MO [L]
Rep. United Assn. of Journeymen & Apprentices of the Plumbing & Pipe Fitting Industry
Michael D. Sides, XL Global Asset Protection Services, FL [I]
Mark Swerdin, Zurich Services Corporation, NY [I]
David Waite, FM Approvals, MA [I]

Alternates

Mark E. Agar, Fire Equipment Company Inc., MI [IM]
(Alt. to N. W. Makowka)
Charles Berra, Sprinkler Fitters LU 268 JAC, MO [L]
(Alt. to K. R. Schneider)
Donald Brighenti, Tyco/SimplexGrinnell, MA [M]
(Alt. to L. L. Leimer)
Charles E. Brockett, XL Global Asset Protection Services, TX [I]
(Alt. to M. D. Sides)
Michael Earl Dillon, Dillon Consulting Engineers, Inc., CA [SE]
(Alt. to R. A. Hall)
Scott M. Golly, Hughes Associates/RJA Group, MD [SE]
(Alt. to M. H. Reiss)
Michael A. Henke, Potter Electric Signal Company, MO [M]
(Alt. to D. L. Royse)
Robert Kasiski, FM Global, MA [I]
(Alt. to D. Waite)
Jack McNamara, Bosch Security Systems, NY [M]
(Alt. to J. W. Mottorn, II)
David E. Mills, UL LLC, IL [RT]
(Alt. to J. L. Parssinen)
David M. Nelson, Volunteer Technology Systems, TN [IM]
(Alt. to O. E. Morphew, Jr.)
Richard S. Pawlish, Aon/Schirmer Engineering Corporation, IL [I]
(Alt. to D. J. O'Connor)
Brian E. Swanick, Siemens Industry, Inc., NJ [M]
(Alt. to R. D. Ouimette)
Jerry Trotter, City of Henderson, NV [E]
(Alt. to L. Nielson)
Michael Yakine, UTC/Kidde-Fenwal, Inc., MA [M]
(Alt. to J. A. Chetelat)

Lee F. Richardson, NFPA Staff Liaison

This list represents the membership at the time the Committee was balloted on the final text of this edition. Since that time, changes in the membership may have occurred. A key to classifications is found at the back of the document.

NOTE: Membership on a committee shall not in and of itself constitute an endorsement of the Association or any document developed by the committee on which the member serves.

Committee Scope: This Committee shall have primary responsibility for documents on the installation and operation of initiating devices for fire alarm and signaling systems.

Technical Committee on Notification Appliances for Fire Alarm and Signaling Systems

David O. Lowrey, *Chair*
City of Boulder Fire Rescue, CO [E]

David E. Becker, *Secretary*
Fire Equipment Service Company, KY [IM]
Rep. National Association of Fire Equipment Distributors

Robert F. Bitter, Honeywell, Inc., MO [M]
Lisa Carr, Christiana Care Health Services, DE [U]
Kristopher Conner, Froula Alarm Systems, Inc., WA [IM]
Michael L. Edwards, U.S. Architect of the Capitol, DC [U]
Raymond A. Grill, Arup, DC [SE]
Daniel M. Grosch, UL LLC, IL [RT]
Jeffrey M. Klein, Honeywell/System Sensor, IL [M]
Rep. Automatic Fire Alarm Association, Inc.
David L. Klepitch, Whitman, Requardt & Associates, LLP, MD [SE]
Doug Kline, Nowak Supply Fire Systems, IN [M]
Rep. Fire Suppression Systems Association
Michael J. Knoras, Jr., Aon Fire Protection Engineering Corporation, GA [I]
Neal W. Krantz, Sr., Krantz Systems & Associates, LLC, MI [IM]
Rep. NFPA Industrial Fire Protection Section
Steven P. Lewis, RFI Enterprises, CA [IM]
Bob D. Morgan, Fort Worth Fire Department, TX [E]
David Newhouse, Gentex Corporation, MI [M]
Rep. National Electrical Manufacturers Association
Maurice M. Pilette, Mechanical Designs Ltd., MA [SE]
Jack Poole, Poole Fire Protection, Inc., KS [SE]
Robert P. Schifiliti, R. P. Schifiliti Associates, Inc., MA [SE]
Michael T. Schmitt, Long Grove Fire Protection District, IL [E]
Daniel L. Seibel, Wolverine Fire Protection Company, MI [IM]
Morris L. Stoops, UTC Fire & Security, FL [M]
Paul R. Strelecki, Siemens Building Technologies, Inc., NJ [M]
John R. Swanson, Minnesota State Fire Marshal Division, MN [E]
Morris Toporek, Environmental Systems Design Inc., IL [SE]

Alternates

Robert F. Accosta, Jr., Arup, NY [SE]
(Alt. to R. A. Grill)
Neal W. Krantz, Jr., Krantz Systems & Associates, LLC, MI [IM]
(Alt. to N. W. Krantz, Sr.)
James Mongeau, Space Age Electronics, Inc., MA [M]
(Alt. to J. M. Klein)
Leon Newsome, Cooper Notification, FL [M]
(Alt. to Da. Newhouse)
Scott E. Panowitz, BFPE International, MD [M]
(Alt. to D. Kline)
Andrew W. Poole, Poole Fire Protection, Inc., KS [SE]
(Alt. to J. Poole)
Casey Lin Quisol, RFI Enterprises, CA [IM]
(Alt. to S. P. Lewis)
Rich Rhome, Froula Alarm Systems, Inc., WA [IM]
(Alt. to K. Conner)
Marlon Erthal Vieira, Siemens Building Technologies, Inc., NJ [M]
(Alt. to P. R. Strelecki)

Lee F. Richardson, NFPA Staff Liaison

This list represents the membership at the time the Committee was balloted on the final text of this edition. Since that time, changes in the membership may have occurred. A key to classifications is found at the back of the document.

NOTE: Membership on a committee shall not in and of itself constitute an endorsement of the Association or any document developed by the committee on which the member serves.

Committee Scope: This Committee shall have primary responsibility for documents on the installation and operation of notification appliances for fire alarm and signaling systems.

Technical Committee on Protected Premises Fire Alarm and Signaling Systems

Merton W. Bunker, Jr., *Chair*
U.S. Department of State, VA [U]

Leonard Belliveau, Jr., *Secretary*
Hughes Associates/RJA Group, RI [SE]

Brian Ballou, Town of South Kingstown, RI [E]
Scott Barrett, Commercial Wireless Systems International (CWSI), FL [M]
James G. Bisker, U.S. Department of Energy, DC [U]
David J. Burkhart, Code Consultants, Inc., MO [SE]
Anthony J. Capowski, Tyco/SimplexGrinnell, MA [M]
Paul F. Crowley, FM Approvals, MA [I]
Keith W. Dix, West Metro Fire Department, CO [E]
Diane P. Doliber, Consultant, NC [SE]
Laura E. Doyle, U.S. General Services Administration, DC [U]
Thomas P. Hammerberg, Automatic Fire Alarm Association, Inc., GA [M]
Rep. Automatic Fire Alarm Association, Inc.
Scott D. Harris, AFA Protective Systems, Inc., NY [IM]
Mark D. Hayes, Aon Fire Protection Engineering Corporation, TX [I]
William K. Hopple, Hopple & Company, CA [IM]
Rep. California Automatic Fire Alarm Association, Inc.
Daniel J. Horon, CADgraphics, Incorporated, MN [M]
Vic Humm, Vic Humm & Associates, TN [SE]
Thomas E. Kuhta, Willis Corporation, NJ [I]
Peter Leszczak, U.S. Department of Veterans Affairs, CT [U]
Scott T. Martorano, The Viking Corporation, MI [M]
Rep. National Fire Sprinkler Association
Jebediah J. Novak, Cedar Rapids Electrical JATC, IA [L]
Rep. International Brotherhood of Electrical Workers
John R. Olenick, Vector Security Inc., MD [IM]
Rep. Central Station Alarm Association
Kurt A. Ruchala, FIREPRO Incorporated, MA [SE]
Lawrence J. Shudak, UL LLC, IL [RT]
Donald Struck, Siemens Fire Safety, NJ [M]
Sagiv Weiss-Ishai, San Francisco Fire Department, CA [E]
Carl F. Willms, Fire Safety Consultants, Inc., NJ [SE]

Alternates

Donald C. Birchler, FP&C Consultants, Inc., MO [SE]
(Alt. to V. Humm)
Richard L. Carter, Siemens Industry, Inc., NJ [M]
(Alt. to D. Struck)
Shane M. Clary, Bay Alarm Company, CA [IM]
(Alt. to W. K. Hopple)
Lee C. DeVito, FIREPRO Incorporated, MA [SE]
(Alt. to K. A. Ruchala)
Joshua W. Elvove, Aurora, CO [SE]
(Alt. to C. F. Willms)
Daniel G. Farley, Tyco Fire Protection Products, MA [M]
(Alt. to A. J. Capowski)
Jacob P. Hemke, Code Consultants, Inc., MO [SE]
(Alt. to D. J. Burkhart)
Theodore Ivers, UL LLC, NY [RT]
(Alt. to L. J. Shudak)
Walter J. Kessler, Jr., FM Approvals, MA [I]
(Alt. to P. F. Crowley)
Peter A. Larrimer, U.S. Department of Veterans Affairs, PA [U]
(Alt. to P. Leszczak)
Timothy J. Lawyer, Aon Fire Protection Engineering Corporation, CA [I]
(Alt. to M. D. Hayes)
David J. LeBlanc, Hughes Associates/RJA Group, MA [SE]
(Alt. to L. Belliveau, Jr.)
Andre Lickefett, Minimax GmbH & Co. KG, Germany [M]
(Alt. to S. T. Martorano)
Michael D. Mann, American Professional Services, Inc., OK [IM]
(Alt. to J. R. Olenick)
Joseph Ranaudo, AFA Protective Systems, Inc., NY [IM]
(Alt. to S. D. Harris)
Scott F. Ruland, Fike Corporation, MO [M]
(Voting Alternate)
Jeffery G. Van Keuren, UTC Climate, Controls & Security/Edwards, FL [M]
(Alt. to T. P. Hammerberg)
Brian Woodward, Fire Safety Engineers Inc., NV [SE]
(Alt. to D. P. Doliber)

Lee F. Richardson, NFPA Staff Liaison

This list represents the membership at the time the Committee was balloted on the final text of this edition. Since that time, changes in the membership may have occurred. A key to classifications is found at the back of the document.

NOTE: Membership on a committee shall not in and of itself constitute an endorsement of the Association or any document developed by the committee on which the member serves.

Committee Scope: This Committee shall have primary responsibility for documents on the installation and operation of protected premises fire alarm and signaling systems, including their interconnection with initiating devices, notification appliances, and other related building control equipment, within the protected premises.

Technical Committee on Public Emergency Reporting Systems

Jeffrey G. Knight, *Chair*
City of Newton Fire Department, MA [U]
Rep. International Municipal Signal Association

Leo F. Martin, Jr., *Secretary*
Martin Electrical Code Consultants, MA [SE]

Douglas M. Aiken, Lakes Region Mutual Fire Aid, NH [E]
George W. Allen, R. B. Allen Company, Inc., NH [IM]
Anthony W. Cole, Wal-Mart Stores, Inc., CA [U]
Daniel R. Dinwiddie, L W Bills Company, MA [M]
Romeo G. Dupuis, Edith Norse Rogers Memorial VA Hospital, MA [U]
Sidney M. Earley, TLC Systems, MA [IM]
John K. Guhl, California State Fire Marshal, CA [E]
Rep. International Association of Fire Chiefs
Paul T. Kahle, Code Consultants, Inc., MO [SE]
Robert E. Lapham, Signal Communications Corporation, MA [M]
Robert Malanga, Fire and Risk Engineering, NJ [SE]
Rep. Fairmount Fire Company No. 1
Max McLeod, Siemens Industry, Inc., AL [M]
Robert E. Myers, MC Dean, VA [IM]
Isa Y. Saah, The Protection Engineering Group, VA [SE]
Deborah L. Shaner, Shaner Life Safety, CO [SE]
Stephen Smith, Advanced Signal Corporation, MA [IM]
Cindy G. Tate, Fort A.P. Hill Fire & Emergency Services, VA [U]
Christopher Willms, Sea Girt Fire Company #1, NJ [E]

Alternates

Dwayne Griffith, Wal-Mart Security Services, AR [U]
(Alt. to A. W. Cole)
Charles E. Hahl, The Protection Engineering Group, Inc., VA [SE]
(Alt. to I. Y. Saah)
Nathaniel M. Johnson, City of Laconia Fire Department, NH [U]
(Alt. to J. G. Knight)
Gregory D. Lapin, King-Fisher Company, Inc., MA [M]
(Alt. to King-Fisher Rep.)
Gene Monaco, Monaco Enterprises, Inc., WA [M]
(Voting Alternate)
Leo J. Watts, Signal Communications Corporation, MA [M]
(Alt. to R. E. Lapham)
Michael S. White, Siemens Industry, Inc., OH [M]
(Alt. to M. McLeod)

Lee F. Richardson, NFPA Staff Liaison

This list represents the membership at the time the Committee was balloted on the final text of this edition. Since that time, changes in the membership may have occurred. A key to classifications is found at the back of the document.

NOTE: Membership on a committee shall not in and of itself constitute an endorsement of the Association or any document developed by the committee on which the member serves.

Committee Scope: This Committee shall have primary responsibility for documents on the proper configuration, performance, installation, and operation of public emergency alarm reporting systems and auxiliary alarm systems. The Committee scope shall include systems that use a communication infrastructure that is publicly owned, operated, and controlled. Reporting of alarms by voice over the public switched telephone network utilizing the Universal Emergency Number 9-1-1, or any other telephone number that can be dialed, is outside the scope of this committee.

Technical Committee on Supervising Station Fire Alarm and Signaling Systems

Warren E. Olsen, *Chair*
Fire Safety Consultants, Inc., IL [E]
Rep. Illinois Fire Inspectors Association

Anthony Mucci, *Secretary*
Tyco Integrated Security, FL [M]

Raymond E. Bigelow, Natick, MA [U]
Rep. International Municipal Signal Association
Art Black, Carmel Fire Protection Associates, CA [E]
David A. Blanken, Keltron Corporation, MA [M]
Edward R. Bonifas, Alarm Detection Systems, Inc., IL [IM]
Robert F. Buckley, Signal Communications Corporation, MA [M]
Paul M. Carroll, Central Signal Corporation, MA [M]
Rep. Automatic Fire Alarm Association, Inc.
Thomas F. Connaughton, Intertek Testing Services, NJ [RT]
Lawrence E. Coveny, Chicago Metropolitan Fire Prevention Company, IL [IM]
James S. Crews, Fireman's Fund Insurance Company, GA [I]
Patrick M. Egan, Select Security, PA [IM]
Bob Elliott, FM Approvals, MA [I]
Louis T. Fiore, L. T. Fiore, Inc., NJ [SE]
Xianxu Hu, Insurance Services Office, Inc., NJ [I]
Michael Johnson, Hughes Associates/RJA Group, WA [SE]
Richard Kleinman, AFA Protective Systems Inc., NY [IM]

Scott M. May, Bosch Security Systems, NY [M]
Rep. National Electrical Manufacturers Association
Randall R. McCarver, Telcordia Technologies (Ericsson), NJ [SE]
Gene Monaco, Monaco Enterprises, Inc., WA [M]
Donald C. Pannell, City of Memphis, TN [E]
Isaac I. Papier, Honeywell, Inc., IL [M]
Steven A. Schmit, UL LLC, IL [RT]
Paul F. Silva, Wayne Alarm Systems, Inc., MA [IM]
Rep. Central Station Alarm Association
Sean P. Titus, Fike Corporation, MO [M]
Rep. Fire Suppression Systems Association
Allyn J. Vaughn, JBA Consulting Engineers, NV [SE]

Alternates

Douglas M. Aiken, Lakes Region Mutual Fire Aid, NH [U]
(Alt. to R. E. Bigelow)
Jeffrey A. Betz, AT&T Corporation, NJ [U]
(Alt. to AT&T Rep.)
William J. (Jody) Dwyer, Germantown Fire Department, TN [E]
(Alt. to D. C. Pannell)
Martin J. Farraher, Siemens Industry, Inc., IL [M]
(Alt. to S. P. Titus)
Gordon G. Hope, Jr., Honeywell, Inc., NY [M]
(Alt. to I. I. Papier)
Kenneth Lowery, Insurance Services Office, Inc., GA [I]
(Alt. to X. Hu)
Larry W. Mann, Central Station, Inc., AL [M]
(Alt. to P. M. Carroll)
Arnold E. Miesch, Intertek, MA [RT]
(Alt. to T. F. Connaughton)
Robert Mitchell, Bay Alarm Company, CA [IM]
(Alt. to P. F. Silva)
Paul J. Olson, UL LLC, MN [RT]
(Alt. to S. A. Schmit)
Rodger Reiswig, Tyco/SimplexGrinnell, FL [M]
(Alt. to A. Mucci)
Steven P. Sargent, Keltron Corporation, MA [M]
(Alt. to D. A. Blanken)
Robert V. Scholes, Fireman's Fund Insurance Company, OR [I]
(Alt. to J. S. Crews)
Kevin T. Thommes, Hughes Associates/RJA Group, CA [SE]
(Alt. to M. Johnson)

Lee F. Richardson, NFPA Staff Liaison

This list represents the membership at the time the Committee was balloted on the final text of this edition. Since that time, changes in the membership may have occurred. A key to classifications is found at the back of the document.

NOTE: Membership on a committee shall not in and of itself constitute an endorsement of the Association or any document developed by the committee on which the member serves.

Committee Scope: This Committee shall have primary responsibility for documents on the installation and operation of equipment for the transmission and receipt of signals from a protected premises to a supervising station, including the supervising station facilities.

Technical Committee on Testing and Maintenance of Fire Alarm and Signaling Systems

J. Jeffrey Moore, *Chair*
Hughes Associates/RJA Group, OH [SE]

Mark L. Rochholz, *Secretary*
Aon Fire Protection Engineering Corporation, IL [I]

Chad E. Beebe, ASHE - AHA, WA [U]
Kevin J. Breen, Breen Electrical Contractors Inc., NY [IM]
Rep. Independent Electrical Contractors, Inc.
Charles E. Brockett, XL Global Asset Protection Services, TX [I]
Steve Carter, Orr Protection Systems, Inc., KY [M]
Rep. Fire Suppression Systems Association
Richard L. Carter, Siemens Industry, Inc., NJ [M]
Louis Chavez, UL LLC, IL [RT]
Scott D. Corrin, University of California-Riverside, CA [U]
Rick Heffernan, SDi, NJ [M]
Rep. National Electrical Manufacturers Association
Herbert B. Hurst, Jr., Savannah River Nuclear Solutions, LLC, SC [U]
Robert H. Kelly, Fire Defense Equipment Company, Inc., MI [IM]
John Kelly, Washington DC Fire & EMS Department, MD [E]
J. David Kerr, Plano Fire Department, TX [E]
Rep. NFPA Fire Service Section
David E. Kipley, AREVA NP, Inc., IL [U]
Rep. Edison Electric Institute
E. J. Kleintop, Tyco/SimplexGrinnell, PA [M]
Chuck Koval, U.S. General Services Administration, WA [U]
Peter A. Larrimer, U.S. Department of Veterans Affairs, PA [U]
James Murphy, Vector Security Inc., PA [IM]
Rep. Central Station Alarm Association
Robert Powers, East Providence Fire Department, RI [E]
Michael J. Reeser, Santa Rosa Fire Equipment Service Inc., CA [IM]
Rep. California Automatic Fire Alarm Association Inc.
Joe Scibetta, BuildingReports, GA [M]
George E. Seymour, Montgomery, TX [IM]
Rep. National Association of Fire Equipment Distributors
Derek Shackley, Pacific Auxiliary Fire Alarm, CA [M]
Rep. Automatic Fire Alarm Association, Inc.
Michael J. Slattery, AFA Protective Systems, Inc., MA [IM]
Timothy M. Soverino, Nantucket, MA [U]
Rep. International Municipal Signal Association
Frank L. Van Overmeiren, FP&C Consultants, Inc., IN [SE]

Alternates

Timothy E. Adams, American Society for Healthcare Engineering, IN [U]
(Alt. to C. E. Beebe)
Scott D. Carlson, Fire Defense Equipment Company, Inc., MI [IM]
(Alt. to R. H. Kelly)
Franklin Chenoweth, Fireline Corporation, MD [M]
(Alt. to S. Carter)
Esteban Cota, Siemens Industry, Inc., IL [M]
(Alt. to R. L. Carter)
Shane Courbier, Gentex Corporation, MI [M]
(Alt. to R. Heffernan)
Ariana Hargrove, City of Rockwall Fire Department, TX [E]
(Alt. to J. D. Kerr)
Vic Humm, Vic Humm & Associates, TN [SE]
(Alt. to F. L. Van Overmeiren)
Bill Isemann, Guardian Fire Protection Services LLC, MD [IM]
(Alt. to G. E. Seymour)
Peter Leszczak, U.S. Department of Veterans Affairs, CT [U]
(Alt. to P. A. Larrimer)
Jay Levy, Hochiki America Corporation, CA [IM]
(Alt. to M. J. Reeser)
James R. Lugar, Hughes Associates/RJA Group, MD [SE]
(Alt. to J. J. Moore)
Chester S. Maciaszek, Savannah River Nuclear Solutions, LLC, SC [U]
(Alt. to H. B. Hurst, Jr.)
Thomas Presnak, UL LLC, IL [RT]
(Alt. to L. Chavez)
Michael D. Sides, XL Global Asset Protection Services, FL [I]
(Alt. to C. E. Brockett)
Charles K. Stormer, U.S. General Services Administration, IL [U]
(Alt. to C. Koval)
Terry L. Victor, Tyco/SimplexGrinnell, MD [M]
(Alt. to E. J. Kleintop)

Lee F. Richardson, NFPA Staff Liaison

This list represents the membership at the time the Committee was balloted on the final text of this edition. Since that time, changes in the membership may have occurred. A key to classifications is found at the back of the document.

NOTE: Membership on a committee shall not in and of itself constitute an endorsement of the Association or any document developed by the committee on which the member serves.

Committee Scope: This Committee shall have primary responsibility for documents and requirements for the inspection, testing, and maintenance of fire alarm and emergency communications systems and associated components, for both new and existing systems. This Committee shall not have responsibility for inspection, testing, and maintenance of single-and multiple-station alarms and household alarm systems.

Contents

NFPA 72

National Fire Alarm and Signaling Code

2016 Edition

IMPORTANT NOTE: This NFPA document is made available for use subject to important notices and legal disclaimers. These notices and disclaimers appear in all publications containing this document and may be found under the heading "Important Notices and Disclaimers Concerning NFPA Standards." They can also be obtained on request from NFPA or viewed at www.nfpa.org/disclaimers.

UPDATES, ALERTS, AND FUTURE EDITIONS: New editions of NFPA codes, standards, recommended practices, and guides (i.e., NFPA Standards) are released on scheduled revision cycles. This edition may be superseded by a later one, or it may be amended outside of its scheduled revision cycle through the issuance of Tentative Interim Amendments (TIAs). An official NFPA Standard at any point in time consists of the current edition of the document, together with any TIAs and Errata in effect. To verify that this document is the current edition or to determine if it has been amended by any TIAs or Errata, please consult the National Fire Codes® Subscription Service or visit the Document Information (DocInfo) pages on the NFPA website at www.nfpa.org/docinfo. In addition to TIAs and Errata, the DocInfo pages also include the option to sign up for Alerts for each document and to be involved in the development of the next edition.

NOTICE: An asterisk (*) following the number or letter designating a paragraph indicates that explanatory material on the paragraph can be found in Annex A.

A reference in brackets following a section or paragraph indicates material that has been extracted from another NFPA document. As an aid to the user, the complete title and edition of the source documents for extracts in mandatory sections of the document are given in Chapter 2 and those for extracts in informational sections are given in Annex H. Extracted text may be edited for consistency and style and may include the revision of internal paragraph references and other references as appropriate. Requests for interpretations or revisions of extracted text shall be sent to the technical committee responsible for the source document.

Information on referenced publications can be found in Chapter 2 and Annex H.

Chapter 1 Administration

1.1 Scope.

1.1.1 *NFPA 72* covers the application, installation, location, performance, inspection, testing, and maintenance of fire alarm systems, supervising station alarm systems, public emergency alarm reporting systems, fire warning equipment and emergency communications systems (ECS), and their components.

1.1.2 The provisions of this chapter apply throughout the Code unless otherwise noted.

1.2* Purpose.

1.2.1 The purpose of this Code is to define the means of signal initiation, transmission, notification, and annunciation; the levels of performance; and the reliability of the various types of fire alarm systems, supervising station alarm systems, public emergency alarm reporting systems, fire warning equipment, emergency communications systems, and their components.

1.2.2 This Code defines the features associated with these systems and also provides information necessary to modify or upgrade an existing system to meet the requirements of a particular system classification.

1.2.3 This Code establishes minimum required levels of performance, extent of redundancy, and quality of installation but does not establish the only methods by which these requirements are to be achieved.

1.2.4* This Code shall not be interpreted to require a level of protection that is greater than that which would otherwise be required by the applicable building or fire code.

1.3 Application.

1.3.1 Alarm systems shall be classified as follows:

(1) Fire alarm systems
  (a) Household fire alarm systems
  (b) Protected premises (local) fire alarm systems
(2) Supervising station alarm systems
  (a) Central station (service) alarm systems
  (b) Remote supervising station alarm systems
  (c) Proprietary supervising station alarm systems
(3) Public emergency alarm reporting systems
  (a) Auxiliary alarm systems — local energy type
  (b) Auxiliary alarm systems — shunt type

1.3.2 Emergency communications systems shall be classified as follows:

(1) One-way emergency communications systems
  (a) Distributed recipient mass notification systems
  (b) In-building fire emergency voice/alarm communications systems
  (c) In-building mass notification systems
  (d) Wide area mass notification systems
(2) Two-way emergency communications systems
  (a) In-building emergency communications systems

1.3.3 Any reference or implied reference to a particular type of hardware shall be for the purpose of clarity and shall not be interpreted as an endorsement.

1.3.4 The intent and meaning of the terms used in this Code shall be, unless otherwise defined herein, the same as those of *NFPA 70.*

1.4 Retroactivity.

1.4.1 Unless otherwise noted, it is not intended that the provisions of this document be applied to facilities, equipment, structures, or installations that were existing or approved for construction or installation prior to the effective date of the document.

1.4.2 In those cases where it is determined by the authority having jurisdiction that the existing situation involves a distinct hazard to life or property, retroactive application of the provisions of this document shall be permitted.

1.5 Equivalency.

1.5.1 Nothing in this Code shall prevent the use of systems, methods, devices, or appliances of equivalent or superior quality, strength, fire resistance, effectiveness, durability, and safety over those prescribed by this Code.

1.5.2 Technical documentation shall be submitted to the authority having jurisdiction to demonstrate equivalency.

1.5.3 The systems, methods, devices, or appliances that are found equivalent shall be approved.

1.6 Units and Formulas.

1.6.1 The units of measure in this Code are presented in U.S. Customary Units (inch-pound units).

1.6.2 Where presented, the International System (SI) of Units follow the inch-pound units in parentheses.

1.6.3 Where both systems of units are presented, either system shall be acceptable for satisfying the requirements in this Code.

1.6.4 Where both systems of units are presented, users of this Code shall apply one set of units consistently and shall not alternate between units.

1.6.5* The values presented for measurements in this Code are expressed with a degree of precision appropriate for practical application and enforcement. It is not intended that the application or enforcement of these values be more precise than the precision expressed.

1.6.6 Where extracted text contains values expressed in only one system of units, the values in the extracted text have been retained without conversion to preserve the values established by the responsible technical committee in the source document.

1.7 Code Adoption Requirements. This Code shall be administered and enforced by the authority having jurisdiction designated by the governing authority. *(See Annex E for sample wording for enabling legislation.)*

Chapter 2 Referenced Publications

2.1 General. The documents or portions thereof listed in this chapter are referenced within this Code and shall be considered part of the requirements of this document.

2.2 NFPA Publications. National Fire Protection Association, 1 Batterymarch Park, Quincy, MA 02169-7471.

NFPA 10, *Standard for Portable Fire Extinguishers,* 2013 edition.

NFPA 13, *Standard for the Installation of Sprinkler Systems,* 2016 edition.

NFPA 25, *Standard for the Inspection, Testing, and Maintenance of Water-Based Fire Protection Systems,* 2014 edition.

NFPA 37, *Standard for the Installation and Use of Stationary Combustion Engines and Gas Turbines,* 2015 edition.

NFPA 70®, *National Electrical Code®,* 2014 edition.

NFPA 75, *Standard for the Fire Protection of Information Technology Equipment,* 2016 edition.

NFPA 90A, *Standard for the Installation of Air-Conditioning and Ventilating Systems,* 2015 edition.

NFPA *101*®, *Life Safety Code®,* 2015 edition.

NFPA 110, *Standard for Emergency and Standby Power Systems,* 2016 edition.

NFPA 111, *Standard on Stored Electrical Energy Emergency and Standby Power Systems,* 2016 edition.

NFPA 170, *Standard for Fire Safety and Emergency Symbols,* 2015 edition.

NFPA 601, *Standard for Security Services in Fire Loss Prevention,* 2015 edition.

NFPA 720, *Standard for the Installation of Carbon Monoxide (CO) Detection and Warning Equipment,* 2015 edition.

NFPA 1031, *Standard for Professional Qualifications for Fire Inspector and Plan Examiner,* 2014 edition.

NFPA 1221, *Standard for the Installation, Maintenance, and Use of Emergency Services Communications Systems,* 2016 edition.

NFPA 1600®, Standard on Disaster/Emergency Management and Business Continuity Programs, 2013 edition.

NFPA 1620, *Standard for Pre-Incident Planning,* 2015 edition.

2.3 Other Publications.

2.3.1 ANSI Publications. American National Standards Institute, Inc., 25 West 43rd Street, 4th Floor, New York, NY 10036.

ANSI A-58.1, *Building Code Requirements for Minimum Design Loads in Buildings and Other Structures.*

ANSI S1.4a, *Specifications for Sound Level Meters,* 1985, reaffirmed 2006.

ANSI S3.41, *American National Standard Audible Emergency Evacuation Signal,* 1990, reaffirmed 2008.

ANSI/ASME A17.1/CSA B44–13, *Safety Code for Elevators and Escalators,* 2013.

ANSI/IEEE C2, *National Electrical Safety Code,* 2007.

ANSI/TIA-568-C.3, *Optical Fiber Cabling Components Standard,* June 2008.

ANSI/UL 217, *Standard for Single and Multiple Station Smoke Alarms,* 6th edition, 2006, revised 2012.

ANSI/UL 268, *Standard for Smoke Detectors for Fire Alarm Systems,* 6th edition, 2009.

ANSI/UL 827, *Standard for Central-Station Alarm Services,* 7th edition, 2008, revised 2013.

ANSI/UL 864, *Standard for Control Units and Accessories for Fire Alarm Systems,* 9th edition, 2003, revised 2012.

ANSI/UL 985, *Standard for Household Fire Warning System Units,* 5th edition, 2000, revised 2008.

ANSI/UL 1638, *Visual Signaling Appliances — Private Mode Emergency and General Utility Signaling,* 4th edition, 2001, revised 2013.

ANSI/UL 1730, *Standard for Smoke Detector Monitors and Accessories for Individual Living Units of Multifamily Residences and Hotel/Motel Rooms,* 4th edition, 2006, revised 2012.

ANSI/UL 1971, *Standard for Signaling Devices for the Hearing Impaired,* 3rd edition, 2002, revised 2013.

ANSI/UL 1981, *Central Station Automation Systems,* 2nd edition, 2003, revised 2012.

ANSI/UL 2017, *Standard for General-Purpose Signaling Devices and Systems,* 2nd edition, 2008, revised 2011.

ANSI/UL 2572, *Mass Notification Systems,* 1st edition, 2011, revised 2012 .

ANSI/UL 60950, *Information Technology Equipment — Safety — Part 1: General Requirements,* 2nd edition, 2007, revised 2011.

2.3.2 EIA Publications. Electronic Industries Alliance, 2500 Wilson Boulevard, Arlington, VA 22201-3834.

EIA Tr 41.3, *Telephones.*

2.3.3 IMSA Publication. International Municipal Signal Association, 165 East Union Street, Newark, NY 14513-0539.

"IMSA Official Wire and Cable Specifications," 2012.

2.3.4 ISO Publications. International Organization for Standardization, 1, ch. de la Voie-Creuse, Case postale 56, CH-1211 Geneva 20, Switzerland.

ISO 7731, *Danger signals for public and work places — Auditory danger signals,* 2003 (reconfirmed 2009).

2.3.5 Telcordia Publications. Telcordia Technologies, One Telcordia Drive, Piscataway, NJ 08854.

GR-506-CORE, LATA Switching Systems Generic Requirements: Signaling for Analog Interface, 2006.

GR-909-CORE, Fiber in the Loop Systems Generic Requirements, 2004.

2.3.6 Other Publications.

Merriam-Webster's Collegiate Dictionary, 11th edition, Merriam-Webster, Inc., Springfield, MA, 2003.

2.4 References for Extracts in Mandatory Sections.

NFPA 70®, National Electrical Code®, 2014 edition.

NFPA *101®*, *Life Safety Code®*, 2015 edition.

NFPA 654, *Standard for the Prevention of Fire and Dust Explosions from the Manufacturing, Processing, and Handling of Combustible Particulate Solids,* 2013 edition.

NFPA 720, *Standard for the Installation of Carbon Monoxide (CO) Detection and Warning Equipment,* 2015 edition.

NFPA 1221, *Standard for the Installation, Maintenance, and Use of Emergency Services Communications Systems,* 2016 edition.

NFPA 5000®, Building Construction and Safety Code®, 2015 edition.

Chapter 3 Definitions

3.1 General. The definitions contained in this chapter shall apply to the terms used in this Code. Where terms are not defined in this chapter or within another chapter, they shall be defined using their ordinarily accepted meanings within the context in which they are used. *Merriam-Webster's Collegiate Dictionary,* 11th edition, shall be the source for the ordinarily accepted meaning.

3.2 NFPA Official Definitions.

3.2.1* Approved. Acceptable to the authority having jurisdiction.

3.2.2* Authority Having Jurisdiction (AHJ). An organization, office, or individual responsible for enforcing the requirements of a code or standard, or for approving equipment, materials, an installation, or a procedure.

3.2.3* Code. A standard that is an extensive compilation of provisions covering broad subject matter or that is suitable for adoption into law independently of other codes and standards.

3.2.4 Labeled. Equipment or materials to which has been attached a label, symbol, or other identifying mark of an organization that is acceptable to the authority having jurisdiction and concerned with product evaluation, that maintains periodic inspection of production of labeled equipment or materials, and by whose labeling the manufacturer indicates compliance with appropriate standards or performance in a specified manner.

3.2.5* Listed. Equipment, materials, or services included in a list published by an organization that is acceptable to the authority having jurisdiction and concerned with evaluation of products or services, that maintains periodic inspection of production of listed equipment or materials or periodic evaluation of services, and whose listing states that either the equipment, material, or service meets appropriate designated standards or has been tested and found suitable for a specified purpose.

3.2.6 Shall. Indicates a mandatory requirement.

3.2.7 Should. Indicates a recommendation or that which is advised but not required.

3.3 General Definitions.

3.3.1 Accessible (as applied to equipment). Admitting close approach; not guarded by locked doors, elevation, or other effective means. [**70:**100] (SIG-FUN)

3.3.2 Accessible (as applied to wiring methods). Capable of being removed or exposed without damaging the building structure or finish or not permanently closed in by the structure or finish of the building. [**70:**100] (SIG-FUN)

3.3.3 Accessible, Readily (Readily Accessible). Capable of being reached quickly for operation, renewal, or inspections without requiring those to whom ready access is requisite to climb over or remove obstacles, or to resort to portable ladders, and so forth. (SIG-FUN)

3.3.4 Accessible Spaces (as applied to detection coverage in Chapter 17). Spaces or concealed areas of construction that can be entered via openable panels, doors hatches, or other readily movable elements (e.g., ceiling tiles). (SIG-IDS)

3.3.5 Acknowledge. To confirm that a message or signal has been received, such as by the pressing of a button or the selection of a software command. (SIG-SSS)

3.3.6* Acoustically Distinguishable Space (ADS). An emergency communications system notification zone, or subdivision thereof, that might be an enclosed or otherwise physically defined space, or that might be distinguished from other spaces because of different acoustical, environmental, or use characteristics, such as reverberation time and ambient sound pressure level. (SIG-NAS)

3.3.7 Active Multiplex System. A multiplexing system in which signaling devices such as transponders are employed to transmit status signals of each initiating device or initiating device circuit within a prescribed time interval so that the lack of receipt of such a signal can be interpreted as a trouble signal. (SIG-SSS)

3.3.8 Addressable Device. A fire alarm system component with discrete identification that can have its status individually identified or that is used to individually control other functions. (SIG-IDS)

3.3.9 Adverse Condition. Any condition occurring in a communications or transmission channel that interferes with the proper transmission or interpretation, or both, of status change signals at the supervising station. *(See also 3.3.253.10, Trouble Signal.)* (SIG-SSS)

3.3.10 Air Sampling–Type Detector. See 3.3.66, Detector.

3.3.11 Alarm. An indication of the existence of a condition that requires immediate response. (SIG-FUN)

3.3.12 Alarm Box.

3.3.12.1 ***Auxiliary Alarm Box.*** An alarm box that can only be operated from one or more remote initiating devices or an auxiliary alarm system used to send an alarm to the communications center. (SIG-PRS)

3.3.12.2 ***Combination Fire Alarm and Guard's Tour Box.*** A manually operated box for separately transmitting a fire alarm signal and a distinctive guard patrol tour supervisory signal. (SIG-IDS)

3.3.12.3 ***Manual Fire Alarm Box.*** A manually operated device used to initiate a fire alarm signal. (SIG-IDS)

3.3.12.4 ***Master Alarm Box.*** A publicly accessible alarm box that can also be operated by one or more remote initiating devices or an auxiliary alarm system used to send an alarm to the communications center. (SIG-PRS)

3.3.12.5 ***Publicly Accessible Alarm Box.*** An enclosure, accessible to the public, housing a manually operated transmitter used to send an alarm to the communications center. (SIG-PRS)

3.3.13 Alarm Repeater System. A device or system for the purpose of automatically retransmitting alarm information received by the alarm processing equipment. (SIG-PRS)

3.3.14 Alarm Service. The service required following the receipt of an alarm signal. (SIG-SSS)

3.3.15 Alarm Signal. See 3.3.253, Signal.

3.3.16 Alarm System. See 3.3.103, Fire Alarm System; 3.3.281, Supervising Station Alarm System; 3.3.211, Public Emergency Alarm Reporting System; 3.3.85.1.2, In-Building Fire Emergency Voice/Alarm Communication System; and 3.3.85.1.3, In-Building Mass Notification System.

3.3.17 Alarm Verification Feature. A feature of automatic fire detection and alarm systems to reduce unwanted alarms wherein smoke detectors report alarm conditions for a minimum period of time, or confirm alarm conditions within a given time period after being reset, in order to be accepted as a valid alarm initiation signal. (SIG-PRO)

3.3.18 Alert Tone. An attention-getting signal to alert occupants of the pending transmission of a voice message. (SIG-PRO)

3.3.19 Analog Initiating Device (Sensor). See 3.3.131, Initiating Device.

3.3.20 Ancillary Functions. Ancillary functions are those non-emergency activations of the fire alarm or mass notification audible, visual, and textual output circuits allowed. Ancillary functions can include general paging, background music, or other non-emergency signals. (SIG-ECS)

3.3.21 Annunciator. A unit containing one or more indicator lamps, alphanumeric displays, or other equivalent means in which each indication provides status information about a circuit, condition, or location. (SIG-FUN)

3.3.22 Apartment Building. A building or portion thereof containing three or more dwelling units with independent cooking and bathroom facilities. (SIG-HOU) [**5000**, 2015]

3.3.23 Audible Notification Appliance. See 3.3.172, Notification Appliance.

3.3.24 Automatic Extinguishing System Supervisory Device. See 3.3.131, Initiating Device.

3.3.25 Automatic Fire Detector. See 3.3.66, Detector.

3.3.26 Automatic Fire Extinguishing or Suppression System Operation Detector. See 3.3.66, Detector.

3.3.27 Autonomous Control Unit (ACU). See 3.3.60, Control Unit.

3.3.28 Auxiliary Alarm System. See 3.3.211, Public Emergency Alarm Reporting System.

3.3.29 Auxiliary Box. See 3.3.12, Alarm Box.

3.3.30* Average Ambient Sound Level. The root mean square, A-weighted, sound pressure level measured over the period of time that any person is present, or a 24-hour period, whichever time period is the lesser. (SIG-NAS)

3.3.31 Beam Construction. See 3.3.38, Ceiling Surfaces.

3.3.32 Building Fire Alarm System. See 3.3.103, Fire Alarm System.

3.3.33 Building Fire Safety Plan. Documentation that provides information on the use of alarms, transmission of alarms, response to alarms, evacuation of immediate area, evacuation of smoke compartment, preparation of floors and building for evacuation and extinguishment of fire. (SIG-ECS)

3.3.34 Carrier. High-frequency energy that can be modulated by voice or signaling impulses. (SIG-SSS)

3.3.35 Carrier System. A means of conveying a number of channels over a single path by modulating each channel on a different carrier frequency and demodulating at the receiving point to restore the signals to their original form. (SIG-SSS)

3.3.36 Ceiling. The upper surface of a space, regardless of height. Areas with a suspended ceiling have two ceilings, one visible from the floor and one above the suspended ceiling. (SIG-IDS)

3.3.36.1 ***Level Ceilings.*** Ceilings that have a slope of less than or equal to 1 in 8. (SIG-IDS)

3.3.36.2 ***Sloping Ceiling.*** A ceiling that has a slope of more than 1 in 8. (SIG-IDS)

3.3.36.3* ***Sloping Peaked-Type Ceiling.*** A ceiling in which the ceiling slopes in two directions from the highest point. Curved or domed ceilings can be considered peaked with the slope figured as the slope of the chord from highest to lowest point. (SIG-IDS)

3.3.36.4* ***Sloping Shed-Type Ceiling.*** A ceiling in which the high point is at one side with the slope extending toward the opposite side. (SIG-IDS)

3.3.37 Ceiling Height. The height from the continuous floor of a room to the continuous ceiling of a room or space. (SIG-IDS)

3.3.38 Ceiling Surfaces.

3.3.38.1 ***Beam Construction.*** Ceilings that have solid structural or solid nonstructural members projecting down from the ceiling surface more than 4 in. (100 mm) and spaced more than 36 in. (910 mm), center to center. (SIG-IDS)

3.3.38.2 ***Girder.*** A support for beams or joists that runs at right angles to the beams or joists. If the top of the girder is within 4 in. (100 mm) of the ceiling, the girder is a factor in determining the number of detectors and is to be considered a beam. If the top of the girder is more than 4 in. (100 mm) from the ceiling, the girder is not a factor in detector location. (SIG-IDS)

3.3.38.3* ***Smooth Ceiling.*** A ceiling surface uninterrupted by continuous projections, such as solid joists, beams, or ducts, extending more than 4 in. (100 mm) below the ceiling surface. (SIG-IDS)

3.3.38.4 ***Solid Joist Construction.*** Ceilings that have solid structural or solid nonstructural members projecting down from the ceiling surface for a distance of more than 4 in. (100 mm) and spaced at intervals of 36 in. (910 mm) or less, center to center. (SIG-IDS)

3.3.39 Central Station. See 3.3.280.1, Central Supervising Station.

3.3.40 Central Station Alarm System. See 3.3.281.1, Central Station Service Alarm System.

3.3.41 Central Station Service. See 3.3.282, Supervising Station Service.

3.3.42 Central Station Service Alarm System. See 3.3.281, Supervising Station Alarm System.

3.3.43 Central Supervising Station. See 3.3.280, Supervising Station.

3.3.44 Channel. A path for voice or signal transmission that uses modulation of light or alternating current within a frequency band. (SIG-SSS)

3.3.44.1 ***Communications Channel.*** A circuit or path connecting a subsidiary station(s) to a supervising station(s) over which signals are carried. (SIG-SSS)

3.3.44.2* ***Radio Channel.*** A band of frequencies of a width sufficient to allow its use for radio communications. (SIG-SSS)

3.3.44.3 ***Transmission Channel.*** A circuit or path connecting transmitters to supervising stations or subsidiary stations on which signals are carried. (SIG-SSS)

3.3.45 Circuit. Either a means of providing power or a connection path between locations *(see 3.3.187)*. (SIG-PRO)

3.3.46 Circuit Interface. See 3.3.136, Interface.

3.3.47 Cloud Chamber Smoke Detection. See 3.3.266, Smoke Detection.

3.3.48* Coded. An audible or visible signal that conveys several discrete bits or units of information. (SIG-NAS)

3.3.49 Combination Detector. See 3.3.66, Detector.

3.3.50 Combination Emergency Communications Systems. See 3.3.86, Emergency Communications Systems — Combination.

3.3.51 Combination Fire Alarm and Guard's Tour Box. See 3.3.12, Alarm Box.

3.3.52 Combination System. See 3.3.103, Fire Alarm System.

3.3.53 Common Talk Mode. See 3.3.291, Talk Mode.

3.3.54* Communications Center. A building or portion of a building that is specifically configured for the primary purpose of providing emergency communications services or public safety answering point (PSAP) services to one or more public safety agencies under the authority or authorities having jurisdiction. [**1221,** 2016] (SIG-PRS)

3.3.55 Communications Channel. See 3.3.44, Channel.

3.3.56 Communications Circuit. Any signaling path of an emergency communications system that carries voice, audio, data, or other signals. (SIG-ECS)

3.3.57 Communications Cloud. The area in the communications path that is supported by providers of communication services not governed under the scope of *NFPA 72* in which signals travel between a protected property and a monitoring station. Depending on the type of transmission that is used, signals can travel on a single defined route or through various routes depending on what is available when the signal is initiated. (SIG-SSS)

3.3.58* Condition. A situation, environmental state, or equipment state of a fire alarm or signaling system. (SIG-FUN)

3.3.58.1 ***Abnormal (Off-Normal) Condition.*** A situation, environmental state, or equipment state that warrants some type of signal, notification, communication, response, action, or service. (SIG-FUN)

3.3.58.1.1* ***Alarm Condition.*** An abnormal condition that poses an immediate threat to life, property, or mission. (SIG-FUN)

3.3.58.1.2* ***Pre-Alarm Condition.*** An abnormal condition that poses a potential threat to life, property, or mission, and time is available for investigation. (SIG-FUN)

3.3.58.1.3* ***Supervisory Condition.*** An abnormal condition in connection with the supervision of other systems, processes, or equipment. (SIG-FUN)

3.3.58.1.4* ***Trouble Condition.*** An abnormal condition in a system due to a fault. (SIG-FUN)

3.3.58.2 ***Normal Condition.*** Circuits, systems, and components are functioning as designed and no abnormal condition exists. (SIG-FUN)

3.3.59 Contiguous Property. See 3.3.203, Property.

3.3.60 Control Unit. A system component that monitors inputs and controls outputs through various types of circuits. (SIG-PRO)

3.3.60.1* ***Autonomous Control Unit (ACU).*** The primary control unit for an in-building mass notification system. (SIG-ECS)

3.3.60.2 ***Emergency Communications Control Unit (ECCU).*** A system capable of sending mass notification messages to individual buildings, zones of buildings, individual outdoor speaker arrays, or zones of outdoor speaker arrays; or a building, multiple buildings, outside areas, or a combination of these. (SIG-ECS)

3.3.60.3 ***Fire Alarm Control Unit.*** See 3.3.100, Fire Alarm Control Unit.

3.3.60.4 ***Wireless Control Unit.*** A component that transmits/receives and processes wireless signals. (SIG-PRO)

3.3.61 Day-Care Home. A building or portion of a building in which more than 3 but not more than 12 clients receive care, maintenance, and supervision, by other than their relative(s) or legal guardian(s), for less than 24 hours per day. [*101,* 2015] (SIG-HOU)

3.3.62 Dedicated Function Fire Alarm Control Unit. See 3.3.100, Fire Alarm Control Unit.

3.3.63 Dedicated Function Fire Alarm System. See 3.3.103, Fire Alarm System.

3.3.64 Deficiency. A condition that interferes with the service or reliability for which the part, system, or equipment was intended. (SIG-TMS)

3.3.65 Delinquency Signal. See 3.3.253, Signal.

3.3.66 Detector. A device suitable for connection to a circuit that has a sensor that responds to a physical stimulus such as gas, heat, or smoke. (SIG-IDS)

3.3.66.1 ***Air Sampling–Type Detector.*** A detector that consists of a piping or tubing distribution network that runs from the detector to the area(s) to be protected. An aspiration fan in the detector housing draws air from the protected area back to the detector through air-sampling ports, piping, or tubing. At the detector, the air is analyzed for fire products. (SIG-IDS)

3.3.66.2 ***Automatic Fire Detector.*** A device designed to detect the presence of a fire signature and to initiate action. For the purpose of this Code, automatic fire detectors are classified as follows: Automatic Fire Extinguishing or Suppression System Operation Detector, Fire–Gas Detector, Heat Detector, Other Fire Detectors, Radiant Energy–Sensing Fire Detector, and Smoke Detector. (SIG-IDS)

3.3.66.3 ***Automatic Fire Extinguishing or Suppression System Operation Detector.*** A device that automatically detects the operation of a fire extinguishing or suppression system by means appropriate to the system employed. (SIG-IDS)

3.3.66.4* ***Combination Detector.*** A device that either responds to more than one of the fire phenomena or employs more than one operating principle to sense one of these phenomena. Typical examples are a combination of a heat detector with a smoke detector or a combination rate-of-rise and fixed-temperature heat detector. This device has listings for each sensing method employed. (SIG-IDS)

3.3.66.5 ***Electrical Conductivity Heat Detector.*** A line-type or spot-type sensing element in which resistance varies as a function of temperature. (SIG-IDS)

3.3.66.6 ***Fire–Gas Detector.*** A device that detects gases produced by a fire. (SIG-IDS)

3.3.66.7* ***Fixed-Temperature Detector.*** A device that responds when its operating element becomes heated to a predetermined level. (SIG-IDS)

3.3.66.8* ***Flame Detector.*** A radiant energy–sensing fire detector that detects the radiant energy emitted by a flame. *(Refer to A.17.8.2.)* (SIG-IDS)

3.3.66.9 ***Gas Detector.*** A device that detects the presence of a specified gas concentration. Gas detectors can be either spot-type or line-type detectors. (SIG-IDS)

3.3.66.10 ***Heat Detector.*** A fire detector that detects either abnormally high temperature or rate-of-temperature rise, or both. (SIG-IDS)

3.3.66.11 ***Line-Type Detector.*** A device in which detection is continuous along a path. Typical examples are rate-of-rise pneumatic tubing detectors, projected beam smoke detectors, and heat-sensitive cable. (SIG-IDS)

3.3.66.12* ***Multi-Criteria Detector.*** A device that contains multiple sensors that separately respond to physical stimulus such as heat, smoke, or fire gases, or employs more than one sensor to sense the same stimulus. This sensor is capable of generating only one alarm signal from the sensors employed in the design either independently or in combination. The sensor output signal is mathematically evaluated to determine when an alarm signal is warranted. The evaluation can be performed either at the detector or at the control unit. This detector has a single listing that establishes the primary function of the detector. (SIG-IDS)

3.3.66.13* ***Multi-Sensor Detector.*** A device that contains multiple sensors that separately respond to physical stimulus such as heat, smoke, or fire gases, or employs more than one sensor to sense the same stimulus. A device capable of generating multiple alarm signals from any one of the sensors employed in the design, independently or in combination. The sensor output signals are mathematically evaluated to determine when an alarm signal is warranted. The evaluation can be performed either at the detector or at the control unit. This device has listings for each sensing method employed. (SIG-IDS)

3.3.66.14 ***Other Fire Detectors.*** Devices that detect a phenomenon other than heat, smoke, flame, or gases produced by a fire. (SIG-IDS)

3.3.66.15 ***Pneumatic Rate-of-Rise Tubing Heat Detector.*** A line-type detector comprising small-diameter tubing, usually copper, that is installed on the ceiling or high on the walls throughout the protected area. The tubing is terminated in a detector unit containing diaphragms and associated contacts set to actuate at a predetermined pressure. The system is sealed except for calibrated vents that compensate for normal changes in temperature. (SIG-IDS)

3.3.66.16 ***Projected Beam–Type Detector.*** A type of photoelectric light obscuration smoke detector wherein the beam spans the protected area. (SIG-IDS)

3.3.66.17 ***Radiant Energy–Sensing Fire Detector.*** A device that detects radiant energy, such as ultraviolet, visible, or infrared, that is emitted as a product of combustion reaction and obeys the laws of optics. (SIG-IDS)

3.3.66.18* ***Rate Compensation Detector.*** A device that responds when the temperature of the air surrounding the device reaches a predetermined level, regardless of the rate-of-temperature rise. (SIG-IDS)

3.3.66.19* ***Rate-of-Rise Detector.*** A device that responds when the temperature rises at a rate exceeding a predetermined value. (SIG-IDS)

3.3.66.20 ***Smoke Detector.*** A device that detects visible or invisible particles of combustion. (SIG-IDS)

3.3.66.21 ***Spark/Ember Detector.*** A radiant energy–sensing fire detector that is designed to detect sparks or embers, or both. These devices are normally intended to operate in dark environments and in the infrared part of the spectrum. (SIG-IDS)

3.3.66.22 ***Spot-Type Detector.*** A device in which the detecting element is concentrated at a particular location. Typical examples are bimetallic detectors, fusible alloy detectors, certain pneumatic rate-of-rise detectors, certain smoke detectors, and thermoelectric detectors. (SIG-IDS)

3.3.67* Device (Class N). A supervised component of a life safety system that communicates with other components of life safety systems and that collects environmental data or performs specific input or output functions necessary to the operation of the life safety system. (SIG-PRO)

3.3.68 Digital Alarm Communicator Receiver (DACR). A system component that accepts and displays signals from digital alarm communicator transmitters (DACTs) sent over the public switched telephone network. (SIG-SSS)

3.3.69 Digital Alarm Communicator System (DACS). A system in which signals are transmitted from a digital alarm communicator transmitter (DACT) located at the protected premises through the public-switched telephone network to a digital alarm communicator receiver (DACR). (SIG-SSS)

3.3.70 Digital Alarm Communicator Transmitter (DACT). A system component at the protected premises to which initiating devices or groups of devices are connected. The DACT seizes the connected telephone line, dials a preselected number to connect to a DACR, and transmits signals indicating a status change of the initiating device. (SIG-SSS)

3.3.71 Digital Alarm Radio Receiver (DARR). A system component composed of two subcomponents: one that receives and decodes radio signals, the other that annunciates the decoded data. These two subcomponents can be coresident at the central station or separated by means of a data transmission channel. (SIG-SSS)

3.3.72 Digital Alarm Radio System (DARS). A system in which signals are transmitted from a digital alarm radio transmitter (DART) located at a protected premises through a radio channel to a digital alarm radio receiver (DARR). (SIG-SSS)

3.3.73 Digital Alarm Radio Transmitter (DART). A system component that is connected to or an integral part of a digital alarm communicator transmitter (DACT) that is used to provide an alternate radio transmission channel. (SIG-SSS)

3.3.74 Display. The visual representation of output data, other than printed copy. (SIG-NAS)

3.3.75 Distributed Recipient Mass Notification System (DRMNS). See 3.3.85, Emergency Communications System.

3.3.76 Dormitory. A building or a space in a building in which group sleeping accommodations are provided for more than 16 persons who are not members of the same family in one room, or a series of closely associated rooms, under joint occupancy and single management, with or without meals, but without individual cooking facilities. [***101***, 2015] (SIG-HOU)

3.3.77* Double Doorway. A single opening that has no intervening wall space or door trim separating the two doors. (SIG-IDS)

3.3.78 Downlink. The radio signal from the base station transmitter to the portable public safety subscriber receiver. (SIG-ECS)

3.3.79 Dwelling Unit. One or more rooms arranged for complete, independent housekeeping purposes with space for eating, living, and sleeping; facilities for cooking; and provisions for sanitation. [**5000,** 2015] (SIG-HOU)

3.3.79.1 ***Multiple Dwelling Unit.*** A building containing three or more dwelling units. (SIG-HOU)

3.3.79.2 ***Single Dwelling Unit.*** A building consisting solely of one dwelling unit. (SIG-HOU)

3.3.80 Effective Masked Threshold. The minimum sound level at which the tone signal is audible in ambient noise. (SIG-NAS)

3.3.81 Electrical Conductivity Heat Detector. See 3.3.66, Detector.

3.3.82* Ember. A particle of solid material that emits radiant energy due either to its temperature or the process of combustion on its surface. *(See also 3.3.272, Spark.)* (SIG-IDS)

3.3.83 Emergency Command Center. See 3.3.87, Emergency Communications System — Emergency Command Center.

3.3.84 Emergency Communications Control Unit (ECCU). See 3.3.60, Control Unit.

3.3.85 Emergency Communications System. A system for the protection of life by indicating the existence of an emergency situation and communicating information necessary to facilitate an appropriate response and action. (SIG-ECS)

3.3.85.1 ***One-Way Emergency Communications System.*** One-way emergency communications systems are intended to broadcast information, in an emergency, to people in one or more specified indoor or outdoor areas. It is intended that emergency messages be conveyed either by audible, visible, or textual means, or any combination thereof. (SIG-ECS)

3.3.85.1.1 ***Distributed Recipient Mass Notification System (DRMNS).*** A distributed recipient mass notification system is a system meant to communicate directly to targeted individuals and groups that might not be in a contiguous area. (SIG-ECS)

3.3.85.1.2 ***In-Building Fire Emergency Voice/Alarm Communications System.*** Dedicated manual or automatic equipment for originating and distributing voice instructions, as well as alert and evacuation signals pertaining to a fire emergency, to the occupants of a building. (SIG-ECS)

3.3.85.1.3 ***In-Building Mass Notification System.*** A system used to provide information and instructions to people in a building(s) or other space using intelligible voice communications and including visible signals, text, graphics, tactile, or other communication methods. (SIG-ECS)

3.3.85.1.4 ***Wide-Area Mass Notification System.*** Wide-area mass notification systems are generally installed to provide real-time information to outdoor areas and could have the capability to communicate with other notification systems provided for a campus, military base, municipality, or similar single or multiple contiguous areas. (SIG-ECS)

3.3.85.2 ***Two-Way Emergency Communications System.*** Two-way emergency communications systems are divided into two categories, those systems that are anticipated to be used by building occupants and those systems that are to be used by fire fighters, police, and other emergency services personnel. Two-way emergency communications systems are used to both exchange information and to communicate information such as, but not limited to, instructions, acknowledgement of receipt of messages, condition of local environment, and condition of persons, and to give assurance that help is on the way. (SIG-ECS)

3.3.86 Emergency Communications Systems — Combination. Various emergency communication systems such as fire alarm, mass notification, fire fighter communications, area of refuge communications, elevator communications, or others that can be served through a single control system or through an interconnection of several control systems. (SIG-ECS)

3.3.87* Emergency Communications System — Emergency Command Center. The room(s) or area(s) staffed during any emergency event by assigned emergency management staff. The room or area contains system communications and control equipment serving one or more buildings where responsible authorities receive information from premises sources or systems or from (higher level) regional or national sources or systems and then disseminate appropriate information to individuals, a building, multiple buildings, outside campus areas, or a combination of these in accordance with the emergency response plan established for the premises. The room or area contains the controls and indicators from which the ECS systems located in the room or area can be manually controlled as required by the emergency response plan and the emergency management coordinator. (SIG-ECS)

3.3.88* Emergency Control Function Interface Device. A listed fire alarm or signaling system component that directly interfaces with the system that operates the emergency control function. (SIG-PRO)

3.3.89* Emergency Control Functions. Building, fire, and emergency control elements or systems that are initiated by the fire alarm or signaling system and either increase the level of life safety for occupants or control the spread of the harmful effects of fire or other dangerous products. (SIG-PRO)

3.3.90* Emergency Response Facility (ERF). A structure or a portion of a structure that houses emergency response agency equipment or personnel for response to alarms. [**1221,** 2016] (SIG-PRS)

3.3.91 Emergency Response Plan. A documented set of actions to address the planning for, management of, and response to natural, technological, and man-made disasters and other emergencies. (SIG-ECS)

3.3.92* Endpoint (Class N). The end of a pathway where a single addressable device or a control unit is connected. (SIG-PRO)

3.3.93* Evacuation. The withdrawal of occupants from a building. (SIG-PRO)

3.3.94 Evacuation Signal. See 3.3.253, Signal.

3.3.95 Executive Software. See 3.3.269, Software.

3.3.96 Exit Marking Audible Notification Appliance. See 3.3.172, Notification Appliance.

3.3.97 False Alarm. See 3.3.304, Unwanted Alarm.

3.3.98 Field of View. The solid cone that extends out from the detector within which the effective sensitivity of the detector is at least 50 percent of its on-axis, listed, or approved sensitivity. (SIG-IDS)

3.3.99 Fire Alarm Control Interface (FACI). See 3.3.136, Interface.

3.3.100* Fire Alarm Control Unit (FACU). A component of the fire alarm system, provided with primary and secondary power sources, which receives signals from initiating devices or other fire alarm control units, and processes these signals to determine part or all of the required fire alarm system output function(s). (SIG-PRO)

3.3.100.1 ***Master Fire Alarm Control Unit.*** A fire alarm control unit that serves the protected premises or portion of the protected premises as a local fire alarm control unit and accepts inputs from other fire alarm control units. (SIG-PRO)

3.3.100.2 ***Protected Premises (Local) Control Unit.*** A fire alarm control unit that serves the protected premises or a portion of the protected premises. (SIG-PRO)

3.3.100.2.1* ***Dedicated Function Fire Alarm Control Unit.*** A protected premises fire alarm control unit that is intended to operate specifically identified emergency control function(s). (SIG-PRO)

3.3.100.2.2 ***Releasing Service Fire Alarm Control Unit.*** A protected premises fire alarm control unit specifically listed for releasing service that is part of a fire suppression system and which provides control outputs to release a fire suppression agent based on either automatic or manual input. (SIG-PRO)

3.3.101 Fire Alarm/Evacuation Signal Tone Generator. A device that produces a fire alarm/evacuation tone upon command. (SIG-PRO)

3.3.102 Fire Alarm Signal. See 3.3.253, Signal.

3.3.103 Fire Alarm System. A system or portion of a combination system that consists of components and circuits arranged to monitor and annunciate the status of fire alarm or supervisory signal-initiating devices and to initiate the appropriate response to those signals. (SIG-FUN)

3.3.103.1* ***Combination System.*** A fire alarm system in which components are used, in whole or in part, in common with a non-fire signaling system. (SIG-PRO)

3.3.103.2 ***Household Fire Alarm System.*** A system of devices that uses a fire alarm control unit to produce an alarm signal in the household for the purpose of notifying the occupants of the presence of a fire so that they will evacuate the premises. (SIG-HOU)

3.3.103.3 ***Municipal Fire Alarm System.*** A public emergency alarm reporting system. (SIG-PRS)

3.3.103.4* ***Protected Premises (Local) Fire Alarm System.*** A fire alarm system located at the protected premises. (SIG-PRO)

3.3.103.4.1 ***Building Fire Alarm System.*** A protected premises fire alarm system that includes any of the features identified in 23.3.3.1 and that serves the general fire alarm needs of a building or buildings and that provides fire department or occupant notification or both. (SIG-PRO)

3.3.103.4.2 ***Dedicated Function Fire Alarm System.*** A protected premises fire alarm system installed specifically to perform emergency control function(s) where a building fire alarm system is not required. (SIG-PRO)

3.3.103.4.3 ***Releasing Fire Alarm System.*** A protected premises fire alarm system that is part of a fire suppression system and/or that provides control inputs to a fire suppression system related to the fire suppression system's sequence of operations and outputs for other signaling and notification. (SIG-PRO)

3.3.104* Fire Command Center. The principal attended or unattended room or area where the status of the detection, alarm communications, control systems, and other emergency systems is displayed and from which the system(s) can be manually controlled. (SIG-ECS)

3.3.105 Fire Extinguisher Electronic Monitoring Device. A device connected to a control unit that monitors the fire extinguisher in accordance with the requirements of NFPA 10. (SIG-IDS)

3.3.106 Fire Warden. A building staff member or a tenant trained to perform assigned duties in the event of a fire emergency. (SIG-PRO)

3.3.107 Fire Warning Equipment. Any detector, alarm, device, or material related to single- and multiple-station alarms or household fire alarm systems. (SIG-HOU)

3.3.108 Fire–Gas Detector. See 3.3.66, Detector.

3.3.109 Fixed-Temperature Detector. See 3.3.66, Detector.

3.3.110 Flame. A body or stream of gaseous material involved in the combustion process and emitting radiant energy at specific wavelength bands determined by the combustion chemistry of the fuel. In most cases, some portion of the emitted radiant energy is visible to the human eye. (SIG-IDS)

3.3.111 Flame Detector. See 3.3.66, Detector.

3.3.112 Flame Detector Sensitivity. The distance along the optical axis of the detector at which the detector can detect a fire of specified size and fuel within a given time frame. (SIG-IDS)

3.3.113 Frequency. Minimum and maximum time between events. (SIG-TMS)

3.3.113.1 ***Weekly Frequency.*** Fifty-two times per year, once per calendar week.

3.3.113.2 ***Monthly Frequency.*** Twelve times per year, once per calendar month.

3.3.113.3 ***Quarterly Frequency.*** Four times per year with a minimum of 2 months, maximum of 4 months.

3.3.113.4 ***Semiannual Frequency.*** Twice per year with a minimum of 4 months, maximum of 8 months.

3.3.113.5 ***Annual Frequency.*** Once per year with a minimum of 9 months, maximum 15 months.

3.3.114 Gateway. A device that is used in the transmission of serial data (digital or analog) from the fire alarm control unit to other building system control units, equipment, or networks and/or from other building system control units to the fire alarm control unit. (SIG-PRO)

3.3.115 Girder. See 3.3.38, Ceiling Surfaces.

3.3.116 Guard's Tour Reporting Station. A device that is manually or automatically initiated to indicate the route being followed and the timing of a guard's tour. (SIG-IDS)

3.3.117 Guard's Tour Supervisory Signal. See 3.3.253, Signal.

3.3.118 Guest Room. An accommodation combining living, sleeping, sanitary, and storage facilities within a compartment. [***101,*** 2015] (SIG-HOU)

3.3.119 Guest Suite. An accommodation with two or more contiguous rooms comprising a compartment, with or without doors between such rooms, that provides living, sleeping, sanitary, and storage facilities. [***101,*** 2015] (SIG-HOU)

3.3.120* Hearing Loss. A full or partial decrease in the ability to detect or comprehend sounds. (SIG-HOU)

3.3.120.1 ***Profound Hearing Loss.*** A hearing threshold of greater than 90 dB.

3.3.121 Heat Alarm. A single- or multiple-station alarm responsive to heat. (SIG-HOU)

3.3.121.1 ***Mechanically Powered, Single-Station Heat Alarm.*** A single-station heat alarm employing a mechanical power source. (SIG-HOU)

3.3.122 Heat Detector. See 3.3.66, Detector.

3.3.123 High Power Speaker Array (HPSA). High power speaker arrays provide capability for voice and tone communications to large outdoor areas. (SIG-ECS)

3.3.124 Hotel. A building or groups of buildings under the same management in which there are sleeping accommodations for more than 16 persons and primarily used by transients for lodging with or without meals. [***101,*** 2015] (SIG-HOU)

3.3.125 Household Fire Alarm System. See 3.3.103, Fire Alarm System.

3.3.126 Hunt Group. A group of associated telephone lines within which an incoming call is automatically routed to an idle (not busy) telephone line for completion. (SIG-SSS)

3.3.127* Identified (as Applied to Equipment). Recognizable as suitable for the specific purpose, function, use, environment, application, and so forth, where described in a particular *Code* requirement. [**70:**100] (SIG-PRS)

3.3.128* Impairment. An abnormal condition, during either a planned or emergency event, where a system, component, or function is inoperable. (SIG-FUN)

3.3.129 In-Building Mass Notification System. See 3.3.85, Emergency Communications System.

3.3.130* In-writing. Any form of correspondence that can be verified upon request. (SIG-FUN)

3.3.131 Initiating Device. A system component that originates transmission of a change-of-state condition, such as in a smoke detector, manual fire alarm box, or supervisory switch. (SIG-IDS)

3.3.131.1 ***Analog Initiating Device (Sensor).*** An initiating device that transmits a signal indicating varying degrees of condition as contrasted with a conventional initiating device, which can only indicate an on–off condition. (SIG-IDS)

3.3.131.2 ***Automatic Extinguishing System Supervisory Device.*** A device that responds to abnormal conditions that could affect the proper operation of an automatic sprinkler system or other fire extinguishing system(s) or suppression system(s), including, but not limited to, control valves, pressure levels, liquid agent levels and temperatures, pump power and running, engine temperature and overspeed, and room temperature. (SIG-IDS)

3.3.131.3 ***Nonrestorable Initiating Device.*** A device in which the sensing element is designed to be destroyed in the process of operation. (SIG-IDS)

3.3.131.4 ***Restorable Initiating Device.*** A device in which the sensing element is not ordinarily destroyed in the process of operation, whose restoration can be manual or automatic. (SIG-IDS)

3.3.131.5 ***Supervisory Signal Initiating Device.*** An initiating device such as a valve supervisory switch, water level indicator, or low air pressure switch on a dry pipe sprinkler system in which the change of state signals an off-normal condition and its restoration to normal of a fire protection or life safety system; or a need for action in connection with guard tours, fire suppression systems or equipment, or maintenance features of related systems. (SIG-IDS)

3.3.132 Initiating Device Circuit. A circuit to which automatic or manual initiating devices are connected where the signal received does not identify the individual device operated. (SIG-PRO)

3.3.133 Inspection Personnel. See 3.3.190, Personnel.

3.3.134 Intelligibility. The quality or condition of being intelligible. (SIG-NAS)

3.3.135* Intelligible. Capable of being understood; comprehensible; clear. (SIG-NAS)

3.3.136 Interface.

3.3.136.1 ***Circuit Interface.*** A circuit component that interfaces initiating devices or control circuits, or both; notification appliances or circuits, or both; system control outputs; and other signaling line circuits to a signaling line circuit. (SIG-PRO)

3.3.136.1.1* ***Emergency Control Function Interface.*** The interface between the fire alarm system emergency control function interface device and the component controlling the emergency control function. (SIG-PRO)

3.3.136.1.2 ***Signaling Line Circuit Interface.*** A system component that connects a signaling line circuit to any combination of initiating devices, initiating device circuits, notification appliances, notification appliance circuits, system control outputs, and other signaling line circuits. (SIG-PRO)

3.3.136.2* ***Fire Alarm Control Interface.*** The fire alarm control interface coordinates signals to and from the fire alarm system and other systems. (SIG-ECS)

3.3.137 Ionization Smoke Detection. See 3.3.266, Smoke Detection.

3.3.138 Leg Facility. The portion of a communications channel that connects not more than one protected premises to a primary or secondary trunk facility. The leg facility includes the portion of the signal transmission circuit from its point of connection with a trunk facility to the point where it is terminated within the protected premises at one or more transponders. (SIG-SSS)

3.3.139 Level Ceilings. See 3.3.36, Ceiling.

3.3.140 Life Safety Network. A type of combination system that transmits fire safety control data through gateways to other building system control units. (SIG-PRO)

3.3.141 Line-Type Detector. See 3.3.66, Detector.

3.3.142 Living Area. Any normally occupiable space in a residential occupancy, other than sleeping rooms or rooms that are intended for combination sleeping/living, bathrooms, toilet compartments, kitchens, closets, halls, storage or utility spaces, and similar areas. [***101,*** 2015] (SIG-HOU)

3.3.143 Loading Capacity. The maximum number of discrete elements of fire alarm systems permitted to be used in a particular configuration. (SIG-SSS)

3.3.144 Local Energy Type Auxiliary Alarm System. See 3.3.211, Public Emergency Alarm Reporting System.

3.3.145* Local Operating Console (LOC). Equipment used by authorized personnel and emergency responders to activate and operate an in-building mass notification system. (SIG-ECS)

3.3.146 Lodging or Rooming House. A building or portion thereof that does not qualify as a one- or two-family dwelling, that provides sleeping accommodations for a total of 16 or fewer people on a transient or permanent basis, without personal care services, with or without meals, but without separate cooking facilities for individual occupants. [***101,*** 2015] (SIG-HOU)

3.3.147 Loss of Power. The reduction of available voltage at the load below the point at which equipment can function as designed. (SIG-FUN)

3.3.148 Low-Power Radio Transmitter/Transceiver. Any device that communicates with associated control/receiving equipment or other transceivers by low-power radio signals. (SIG-PRO)

3.3.149 Maintenance. Work, including, but not limited to, repair, replacement, and service, performed to ensure that equipment operates properly. (SIG-TMS)

3.3.150 Malicious Alarm. See 3.3.304.1, Unwanted Alarm.

3.3.151* Managed Facilities-Based Voice Network (MFVN). A physical facilities-based network capable of transmitting real time signals with formats unchanged that is managed, operated, and maintained by the service provider to ensure service quality and reliability from the subscriber location to public-switched telephone network (PSTN) interconnection points or other MFVN peer networks. (SIG-SSS)

3.3.152 Manual Fire Alarm Box. See 3.3.12, Alarm Box.

3.3.153* Manufacturer's Published Instructions. Published installation and operating documentation provided for each product or component. The documentation includes directions and necessary information for the intended installation, maintenance, and operation of the product or component. (SIG-TMS)

3.3.154* Mass Notification Priority Mode. The mode of operation whereby all fire alarm occupant notification is superseded by emergency mass notification action. (SIG-ECS)

3.3.155* Mass Notification System. See 3.3.85.1.3, In-Building Mass Notification System. (SIG-PRO)

3.3.156 Master Box. See 3.3.12, Alarm Box.

3.3.157 Master Fire Alarm Control Unit. See 3.3.100, Fire Alarm Control Unit.

3.3.158 Multi-Criteria Detector. See 3.3.66, Detector.

3.3.159 Multiple Dwelling Unit. See 3.3.79, Dwelling Unit.

3.3.160 Multiple-Station Alarm. A single-station alarm capable of being interconnected to one or more additional alarms so that the actuation of one causes the appropriate alarm signal to operate in all interconnected alarms. (SIG-HOU)

3.3.161 Multiple-Station Alarm Device. Two or more single-station alarm devices that can be interconnected so that actuation of one causes all integral or separate audible alarms to operate; or one single-station alarm device having connections to other detectors or to a manual fire alarm box. (SIG-HOU)

3.3.162 Multiplexing. A signaling method characterized by simultaneous or sequential transmission, or both, and reception of multiple signals on a signaling line circuit, a transmission channel, or a communications channel, including means for positively identifying each signal. (SIG-SSS)

3.3.163 Multi-Sensor Detector. See 3.3.66, Detector.

3.3.164 Municipal Fire Alarm Box (Street Box). A publicly accessible alarm box. (See 3.3.12, Alarm Box.)

3.3.165 Municipal Fire Alarm System. See 3.3.103, Fire Alarm System.

3.3.166 Net-Centric Alerting System (NCAS). A net-centric alerting system incorporates web-based management and alert activation application through which all operators and administrators could gain access to the system's capabilities based on the users' permissions and the defined access policy. (SIG-ECS)

3.3.167 Network.

3.3.167.1 ***Wired Network (Public Emergency Alarm Reporting Systems).*** The method of communications used in a public emergency alarm reporting system that consists of two or more points that are connected by physical conductors. (SIG-PRS)

3.3.167.2 ***Wireless Network (Public Emergency Alarm Reporting Systems).*** The method of communications used in a public emergency alarm reporting system that consists of two or more points that are not connected by physical conductors. (SIG-PRS)

3.3.168 Network Architecture. The physical and logical design of a network, and the inherent ability of the design to carry data from one point to another. (SIG-ECS)

3.3.169 Noncontiguous Property. See 3.3.203, Property.

3.3.170* Nonrequired. A system component or group of components that is installed at the option of the owner, and is not installed due to a building or fire code requirement. (SIG-FUN)

3.3.171 Nonrestorable Initiating Device. See 3.3.131, Initiating Device.

3.3.172 Notification Appliance. A fire alarm system component such as a bell, horn, speaker, light, or text display that provides audible, tactile, or visible outputs, or any combination thereof. (SIG-NAS)

3.3.172.1 ***Audible Notification Appliance.*** A notification appliance that alerts by the sense of hearing. (SIG-NAS)

3.3.172.1.1 ***Exit Marking Audible Notification Appliance.*** An audible notification appliance that marks building exits and areas of refuge by the sense of hearing for the purpose of evacuation or relocation. (SIG-NAS)

3.3.172.1.2* ***Textual Audible Notification Appliance.*** A notification appliance that conveys a stream of audible information. (SIG-NAS)

3.3.172.2 ***Tactile Notification Appliance.*** A notification appliance that alerts by the sense of touch or vibration. (SIG-NAS)

3.3.172.3 ***Visible Notification Appliance.*** A notification appliance that alerts by the sense of sight. (SIG-NAS)

3.3.172.3.1* ***Textual Visible Notification Appliance.*** A notification appliance that conveys a stream of visible information that displays an alphanumeric or pictorial message. (SIG-NAS)

3.3.173 Notification Appliance Circuit. A circuit or path directly connected to a notification appliance(s). (SIG-PRO)

3.3.174 Notification Zone. See 3.3.317, Zone.

3.3.175 Nuisance Alarm. See 3.3.304.2, Unwanted Alarm.

3.3.176* Occupiable. A room or enclosed space designed for human occupancy. (SIG-FUN)

3.3.177 Occupiable Area. An area of a facility occupied by people on a regular basis. (SIG-FUN)

3.3.178* Octave Band. The bandwidth of a filter that comprises a frequency range of a factor of 2. (SIG-NAS)

3.3.178.1 ***One-Third Octave Band.*** The bandwidth of a filter that comprises a frequency range of a factor of $2^{1/3}$. (SIG-NAS)

3.3.179 Off-Hook. To make connection with the public-switched telephone network in preparation for dialing a telephone number. (SIG-SSS)

3.3.180 One-Third Octave Band. See 3.3.178, Octave Band.

3.3.181 One-Way Emergency Communications System. See 3.3.85, Emergency Communications System.

3.3.182 On-Hook. To disconnect from the public-switched telephone network. (SIG-SSS)

3.3.183 Operating Mode.

3.3.183.1 ***Private Operating Mode.*** Audible or visible signaling only to those persons directly concerned with the implementation and direction of emergency action initiation and procedure in the area protected by the fire alarm system. (SIG-NAS)

3.3.183.2 ***Public Operating Mode.*** Audible or visible signaling to occupants or inhabitants of the area protected by the fire alarm system. (SIG-NAS)

3.3.184 Other Fire Detectors. See 3.3.66, Detector.

3.3.185* Ownership. Any property or building or its contents under legal control by the occupant, by contract, or by holding of a title or deed. (SIG-SSS)

3.3.186 Paging System. A system intended to page one or more persons by such means as voice over loudspeaker, coded audible signals or visible signals, or lamp annunciators. (SIG-PRO)

3.3.187 Path (Pathways). Any circuit, conductor, optic fiber, radio carrier, or other means connecting two or more locations. (SIG-PRO)

3.3.188 Pathway Survivability. The ability of any conductor, optic fiber, radio carrier, or other means for transmitting system information to remain operational during fire conditions. (SIG-ECS)

3.3.189 Permanent Visual Record (Recording). An immediately readable, not easily alterable, print, slash, or punch record of all occurrences of status change. (SIG-SSS)

3.3.190 Personnel.

3.3.190.1 ***Inspection Personnel.*** Individuals who conduct a visual examination of a system or portion thereof to verify that it appears to be in operating condition, in proper location, and is free of physical damage or conditions that impair operation. (SIG-TMS)

3.3.190.2 ***Service Personnel.*** Individuals who perform those procedures, adjustments, replacement of components, system programming, and maintenance as described in the manufacturer's service instructions that can affect any aspect of the performance of the system. (SIG-TMS)

3.3.190.3 ***System Designer.*** Individual responsible for the development of fire alarm and signaling system plans and specifications in accordance with this Code. (SIG-FUN)

3.3.190.4 ***System Installer.*** Individual responsible for the proper installation of fire alarm and signaling systems in accordance with plans, specifications, and manufacturer's requirements. (SIG-FUN)

3.3.190.5 ***Testing Personnel.*** Individuals who perform procedures used to determine the status of a system as intended by conducting acceptance, reacceptance, or periodic physical checks on systems. (SIG-TMS)

3.3.191 Photoelectric Light Obscuration Smoke Detection. See 3.3.266, Smoke Detection.

3.3.192 Photoelectric Light-Scattering Smoke Detection. See 3.3.266, Smoke Detection.

3.3.193 Plant. One or more buildings under the same ownership or control on a single property. (SIG-SSS)

3.3.194 Pneumatic Rate-of-Rise Tubing Heat Detector. See 3.3.66, Detector.

3.3.195 Positive Alarm Sequence. An automatic sequence that results in an alarm signal, even when manually delayed for investigation, unless the system is reset. (SIG-PRO)

3.3.196 Power Supply. A source of electrical operating power, including the circuits and terminations connecting it to the dependent system components. (SIG-FUN)

3.3.197 Primary Battery (Dry Cell). A nonrechargeable battery requiring periodic replacement. (SIG-FUN)

3.3.198 Primary Trunk Facility. That part of a transmission channel connecting all leg facilities to a supervising or subsidiary station. (SIG-SSS)

3.3.199 Prime Contractor. The one company contractually responsible for providing central station services to a subscriber as required by this Code. The prime contractor can be either a listed central station or a listed alarm service–local company. (SIG-SSS)

3.3.200 Private Operating Mode. See 3.3.183, Operating Mode.

3.3.201 Profound Hearing Loss. See 3.3.120, Hearing Loss.

3.3.202 Projected Beam–Type Detector. See 3.3.66, Detector.

3.3.203 Property.

3.3.203.1 ***Contiguous Property.*** A single-owner or single-user protected premises on a continuous plot of ground, including any buildings thereon, that is not separated by a public thoroughfare, transportation right-of-way, property owned or used by others, or body of water not under the same ownership. (SIG-SSS)

3.3.203.2 ***Noncontiguous Property.*** An owner- or user-protected premises where two or more protected premises, controlled by the same owner or user, are separated by a public thoroughfare, body of water, transportation right-of-way, or property owned or used by others. (SIG-SSS)

3.3.204 Proprietary Supervising Station. See 3.3.280, Supervising Station.

3.3.205 Proprietary Supervising Station Alarm System. See 3.3.281, Supervising Station Alarm System.

3.3.206 Proprietary Supervising Station Service. See 3.3.282, Supervising Station Service.

3.3.207 Protected Premises. The physical location protected by a fire alarm system. (SIG-PRO)

3.3.208 Protected Premises (Local) Control Unit. See 3.3.100, Fire Alarm Control Unit.

3.3.209 Protected Premises (Local) Fire Alarm System. See 3.3.103, Fire Alarm System.

3.3.210 Public Address System. An electronic amplification system with a mixer, amplifier, and loudspeakers, used to reinforce a given sound and distributing the "sound" to the general public around a building. (SIG-ECS)

3.3.211 Public Emergency Alarm Reporting System. A system of alarm-initiating devices, transmitting and receiving equipment, and communication infrastructure (other than a public telephone network) used to communicate with the communications center to provide any combination of manual or auxiliary alarm service. (SIG-PRS)

3.3.211.1* ***Auxiliary Alarm System.*** A protected premises fire alarm system or other emergency system at the protected premises and the system used to connect the protected premises system to a public emergency alarm reporting system for transmitting an alarm to the communications center. (SIG-PRS)

3.3.211.1.1 ***Local Energy Type Auxiliary Alarm System.*** An auxiliary system that employs a locally complete arrangement of parts, initiating devices, relays, power supply, and associated components to automatically activate a master box or auxiliary box over circuits that are electrically isolated from the public emergency alarm reporting system circuits. (SIG-PRS)

3.3.211.1.2 ***Shunt-Type Auxiliary Alarm System.*** An auxiliary system electrically connected to the public emergency alarm reporting system extending a public emergency alarm reporting circuit to interconnect initiating devices within a protected premises, which, when operated, opens the public emergency alarm reporting circuit shunted around the trip coil of the master box or auxiliary box. The master box or auxiliary box is thereupon energized to start transmission without any assistance from a local source of power. (SIG-PRS)

3.3.211.2 ***Type A Public Emergency Alarm Reporting System.*** A system in which an alarm from an alarm box is received and is retransmitted to an emergency response facility either manually or automatically. (SIG-PRS)

3.3.211.3 ***Type B Public Emergency Alarm Reporting System.*** A system in which an alarm from an alarm box is automatically transmitted to an emergency response facility and, if used, is transmitted to supplementary alerting devices. (SIG-PRS)

3.3.212 Public Operating Mode. See 3.3.183, Operating Mode.

3.3.213 Public Safety Agency. A fire, emergency medical services, or law enforcement agency. (SIG-ECS)

3.3.214 Public Safety Radio Enhancement System. A system installed to assure the effective operation of radio communication systems used by fire, emergency medical services, or law enforcement agencies. (SIG-ECS)

3.3.215 Public Safety Radio System. A radio communication system used by fire, emergency medical services, or law enforcement agencies. (SIG-ECS)

3.3.216 Public Switched Telephone Network. See 3.3.287, Switched Telephone Network.

3.3.217 Publicly Accessible Fire Alarm Box. See 3.3.12, Fire Alarm Box.

3.3.218* Qualified. A competent and capable person or company that has met the requirements and training for a given field acceptable to the authority having jurisdiction. (SIG-TMS)

3.3.219 Radiant Energy–Sensing Fire Detector. See 3.3.66, Detector.

3.3.220 Radio Alarm Repeater Station Receiver (RARSR). A system component that receives radio signals and resides at a repeater station that is located at a remote receiving location. (SIG-SSS)

3.3.221 Radio Alarm Supervising Station Receiver (RASSR). A system component that receives data and annunciates that data at the supervising station. (SIG-SSS)

3.3.222 Radio Alarm System (RAS). A system in which signals are transmitted from a radio alarm transmitter (RAT) located at a protected premises through a radio channel to two or more radio alarm repeater station receivers (RARSR) and that are annunciated by a radio alarm supervising station receiver (RASSR) located at the supervising station. (SIG-SSS)

3.3.223 Radio Alarm Transmitter (RAT). A system component at the protected premises to which initiating devices or groups of devices are connected that transmits signals indicating a status change of the initiating devices. (SIG-SSS)

3.3.224 Radio Channel. See 3.3.44, Channel.

3.3.225* Radio Frequency. The number of electromagnetic wave frequency cycles transmitted by a radio in 1 second. [**1221,** 2016] (SIG-PRS)

3.3.226 Rate Compensation Detector. See 3.3.66, Detector.

3.3.227 Rate-of-Rise Detector. See 3.3.66, Detector.

3.3.228 Record Drawings. Drawings (as-built) that document the location of all devices, appliances, wiring sequences, wiring methods, and connections of the components of the system as installed. (SIG-FUN)

3.3.229 Record of Completion. A document that acknowledges the features of installation, operation (performance), service, and equipment with representation by the property owner, system installer, system supplier, service organization, and the authority having jurisdiction. (SIG-FUN)

3.3.230 Releasing Fire Alarm System. See 3.3.103, Fire Alarm System.

3.3.231 Releasing Service Fire Alarm Control Unit. See 3.3.100, Fire Alarm Control Unit.

3.3.232 Relocation. The movement of occupants from a fire zone to a safe area within the same building. (SIG-PRO)

3.3.233 Remote Supervising Station. See 3.3.280, Supervising Station.

3.3.234 Remote Supervising Station Alarm System. See 3.3.281, Supervising Station Alarm System.

3.3.235 Remote Supervising Station Service. See 3.3.282, Supervising Station Service.

3.3.236 Repeater Station. The location of the equipment needed to relay signals between supervising stations, subsidiary stations, and protected premises. (SIG-SSS)

3.3.237 Reset. A control function that attempts to return a system or device to its normal, nonalarm state. (SIG-FUN)

3.3.238 Residential Board and Care Occupancy. An occupancy used for lodging and boarding of four or more residents, not related by blood or marriage to the owners or operators, for the purpose of providing personal care services. [***101,*** 2015] (SIG-HOU)

3.3.239 Residential Occupancy. An occupancy that provides sleeping accommodations for purposes other than health care or detention and correctional. [***101,*** 2015] (SIG-HOU)

3.3.240* Response. Actions performed upon the receipt of a signal. (SIG-FUN)

3.3.240.1* ***Alarm Response.*** The response to the receipt of an alarm signal. (SIG-FUN)

3.3.240.2* ***Pre-Alarm Response.*** The response to the receipt of a pre-alarm signal. (SIG-FUN)

3.3.240.3* ***Supervisory Response.*** The response to the receipt of a supervisory signal. (SIG-FUN)

3.3.240.4* ***Trouble Response.*** The response to the receipt of a trouble signal. (SIG-FUN)

3.3.241 Response Time Index (RTI). A numerical value that represents the thermal response sensitivity of the sensing element in a heat detector, sprinkler, or other heat-sensing fire detection device to the fire environment in terms of gas temperature and velocity versus time. *(See B.3.3.3.7.)* (SIG-IDS)

3.3.242 Restorable Initiating Device. See 3.3.131, Initiating Device.

3.3.243 Risk Analysis. A process to characterize the likelihood, vulnerability, and magnitude of incidents associated with natural, technological, and manmade disasters and other emergencies that address scenarios of concern, their probability, and their potential consequences. (SIG-ECS)

3.3.244 Runner. A person other than the required number of operators on duty at central, supervising, or runner stations (or otherwise in contact with these stations) available for prompt dispatching, when necessary, to the protected premises. (SIG-SSS)

3.3.245 Runner Service. The service provided by a runner at the protected premises, including restoration, resetting, and silencing of all equipment transmitting fire alarm or supervisory or trouble signals to an off-premises location. (SIG-SSS)

3.3.246 Secondary Trunk Facility. That part of a transmission channel connecting two or more, but fewer than all, leg facilities to a primary trunk facility. (SIG-SSS)

3.3.247 Selective Talk Mode. See 3.3.291, Talk Mode.

3.3.248 Separate Sleeping Area. The area of a dwelling unit where the bedrooms or sleeping rooms are located. [**720,** 2015] (SIG-HOU)

3.3.249 Service Personnel. See 3.3.190, Personnel.

3.3.250 Shapes of Ceilings. The shapes of ceilings can be classified as sloping or smooth. (SIG-IDS)

3.3.251* Shop Drawings. Documents that provide information pertaining to the system necessary for installation of a fire alarm and/or signaling system. (SIG-FUN)

3.3.252 Shunt-Type Auxiliary Alarm System. See 3.3.211, Public Emergency Alarm Reporting System.

3.3.253* Signal. An indication of a condition communicated by electrical, visible, audible, wireless, or other means. (SIG-FUN)

3.3.253.1* ***Alarm Signal.*** A signal that results from the manual or automatic detection of an alarm condition. (SIG-FUN)

3.3.253.2 ***Carbon Monoxide Alarm Signal.*** A signal indicating a concentration of carbon monoxide at or above the alarm threshold that could pose a risk to the life safety of the occupants and that requires immediate action. [**720,** 2015] (SIG-FUN)

3.3.253.3 ***Delinquency Signal.*** A signal indicating a supervisory condition and the need for action in connection with the supervision of guards or system attendants. (SIG-PRO)

3.3.253.4 ***Evacuation Signal.*** A distinctive alarm signal intended to be recognized by the occupants as requiring evacuation of the building. (SIG-PRO)

3.3.253.5* ***Fire Alarm Signal.*** A signal that results from the manual or automatic detection of a fire alarm condition. (SIG-FUN)

3.3.253.6* ***Guard's Tour Supervisory Signal.*** A signal generated when a guard on patrol has activated a guard's tour reporting station. (SIG-PRO)

3.3.253.7* ***Pre-Alarm Signal.*** A signal that results from the detection of a pre-alarm condition. (SIG-FUN)

3.3.253.8 ***Restoration Signal.*** A signal that results from the return to normal condition of an initiating device, system element, or system. (SIG-FUN)

3.3.253.9* ***Supervisory Signal.*** A signal that results from the detection of a supervisory condition. (SIG-FUN)

3.3.253.10* ***Trouble Signal.*** A signal that results from the detection of a trouble condition. (SIG-FUN)

3.3.254 Signal Transmission Sequence. A DACT that obtains dial tone, dials the number(s) of the DACR, obtains verification that the DACR is ready to receive signals, transmits the signals, and receives acknowledgment that the DACR has accepted that signal before disconnecting (going on-hook). (SIG-SSS)

3.3.255 Signaling Line Circuit. A circuit path between any combination of addressable appliances or devices, circuit interfaces, control units, or transmitters over which multiple system input signals or output signals or both are carried. (SIG-PRO)

3.3.256 Signaling Line Circuit Interface. See 3.3.136, Interface.

3.3.257 Signaling Zone. See 3.3.317, Zone.

3.3.258 Single Dwelling Unit. See 3.3.79, Dwelling Unit.

3.3.259 Single-Station Alarm. A detector comprising an assembly that incorporates a sensor, control components, and an alarm notification appliance in one unit operated from a power source either located in the unit or obtained at the point of installation. (SIG-HOU)

3.3.260 Single-Station Alarm Device. An assembly that incorporates the detector, the control equipment, and the alarm-sounding device in one unit operated from a power supply either in the unit or obtained at the point of installation. (SIG-HOU)

3.3.261 Site-Specific Software. See 3.3.269, Software.

3.3.262 Sloping Ceiling. See 3.3.36, Ceiling.

3.3.263 Sloping Peaked-Type Ceiling. See 3.3.36, Ceiling.

3.3.264 Sloping Shed-Type Ceiling. See 3.3.36, Ceiling.

3.3.265 Smoke Alarm. A single or multiple-station alarm responsive to smoke. (SIG-HOU)

3.3.266 Smoke Detection.

3.3.266.1 ***Cloud Chamber Smoke Detection.*** The principle of using an air sample drawn from the protected area into a high-humidity chamber combined with a lowering of chamber pressure to create an environment in which the resultant moisture in the air condenses on any smoke particles present, forming a cloud. The cloud density is measured by a photoelectric principle. The density signal is processed and used to convey an alarm condition when it meets preset criteria. (SIG-IDS)

3.3.266.2* ***Ionization Smoke Detection.*** The principle of using a small amount of radioactive material to ionize the air between two differentially charged electrodes to sense the presence of smoke particles. Smoke particles entering the ionization volume decrease the conductance of the air by reducing ion mobility. The reduced conductance signal is processed and used to convey an alarm condition when it meets preset criteria. (SIG-IDS)

3.3.266.3* ***Photoelectric Light Obscuration Smoke Detection.*** The principle of using a light source and a photosensitive sensor onto which the principal portion of the source emissions is focused. When smoke particles enter the light path, some of the light is scattered and some is absorbed, thereby reducing the light reaching the receiving sensor. The light reduction signal is processed and used to convey an alarm condition when it meets preset criteria. (SIG-IDS)

3.3.266.4* ***Photoelectric Light-Scattering Smoke Detection.*** The principle of using a light source and a photosensitive sensor arranged so that the rays from the light source do not normally fall onto the photosensitive sensor. When smoke particles enter the light path, some of the light is scattered by reflection and refraction onto the sensor. The light signal is processed and used to convey an alarm condition when it meets preset criteria. (SIG-IDS)

3.3.266.5* ***Video Image Smoke Detection (VISD).*** The principle of using automatic analysis of real-time video images to detect the presence of smoke. (SIG-IDS)

3.3.267 Smoke Detector. See 3.3.66, Detector.

3.3.268 Smooth Ceiling. See 3.3.38, Ceiling Surfaces.

3.3.269 Software. Programs, instruments, procedures, data, and the like that are executed by a central processing unit of a product and that influences the functional performance of that product. For the purpose of this Code, software is one of two types: executive software and site-specific software. (SIG-TMS)

3.3.269.1 ***Executive Software.*** Control and supervisory program that manages the execution of all other programs and directly or indirectly causes the required functions of the product to be performed. Executive software is sometimes referred to as firmware, BIOS, or executive program. (SIG-TMS)

3.3.269.2 ***Site-Specific Software.*** Program that is separate from, but controlled by, the executive software that allows inputs, outputs, and system configuration to be selectively defined to meet the needs of a specific installation. Typically it defines the type and quantity of hardware, customized labels, and the specific operating features of a system. (SIG-TMS)

3.3.270 Solid Joist Construction. See 3.3.38, Ceiling Surfaces.

3.3.271 Spacing. A horizontally measured dimension used as a criterion in determining the allowable coverage of devices. (SIG-FUN)

3.3.272* Spark. A moving particle of solid material that emits radiant energy due to either its temperature or the process of combustion on its surface. [**654,** 2013] (SIG-IDS)

3.3.273 Spark/Ember Detector. See 3.3.66, Detector.

3.3.274 Spark/Ember Detector Sensitivity. The number of watts (or the fraction of a watt) of radiant power from a point source radiator, applied as a unit step signal at the wavelength of maximum detector sensitivity, necessary to produce an alarm signal from the detector within the specified response time. (SIG-IDS)

3.3.275 Spot-Type Detector. See 3.3.66, Detector.

3.3.276 Stakeholder. Any individual, group, or organization that might affect, be affected by, or perceive itself to be affected by the risk. (SIG-ECS)

3.3.277 Stratification. The phenomenon where the upward movement of smoke and gases ceases due to the loss of buoyancy. (SIG-IDS)

3.3.278 Subscriber. The recipient of a contractual supervising station signal service(s). In case of multiple, noncontiguous properties having single ownership, the term refers to each protected premises or its local management. (SIG-SSS)

3.3.279 Subsidiary Station. A subsidiary station is a normally unattended location that is remote from the supervising station and is linked by a communications channel(s) to the supervising station. Interconnection of signals on one or more transmission channels from protected premises with a communications channel(s) to the supervising station is performed at this location. (SIG-SSS)

3.3.280 Supervising Station. A facility that receives signals from protected premises fire alarm systems and at which personnel are in attendance at all times to respond to these signals. (SIG-SSS)

3.3.280.1 ***Central Supervising Station.*** A supervising station that is listed for central station service and that also commonly provides less stringent supervising station services such as remote supervising services. (SIG-SSS)

3.3.280.2 ***Proprietary Supervising Station.*** A supervising station under the same ownership as the protected premises fire alarm system(s) that it supervises (monitors) and to which alarm, supervisory, or trouble signals are received and where personnel are in attendance at all times to supervise operation and investigate signals. (SIG-SSS)

3.3.280.3 ***Remote Supervising Station.*** A supervising station to which alarm, supervisory, or trouble signals or any combination of those signals emanating from protected premises fire alarm systems are received and where personnel are in attendance at all times to respond. (SIG-SSS)

3.3.281 Supervising Station Alarm Systems.

3.3.281.1 ***Central Station Service Alarm System.*** A system or group of systems in which the operations of circuits and devices are transmitted automatically to, recorded in, maintained by, and supervised from a listed central station that has competent and experienced servers and operators who, upon receipt of a signal, take such action as required by this Code. Such service is to be controlled and operated by a person, firm, or corporation whose business is the furnishing, maintaining, or monitoring of supervised alarm systems. (SIG-SSS)

3.3.281.2 ***Proprietary Supervising Station Alarm System.*** An installation of an alarm system that serves contiguous and noncontiguous properties, under one ownership, from a proprietary supervising station located at the protected premises, or at one of multiple noncontiguous protected premises, at which trained, competent personnel are in constant attendance. This includes the protected premises fire alarm system(s); proprietary supervising station; power supplies; signal-initiating devices; initiating device circuits; signal notification appliances; equipment for the automatic, permanent visual recording of signals; and equipment for initiating the operation of emergency building control services. (SIG-SSS)

3.3.281.3 ***Remote Supervising Station Alarm System.*** A protected premises fire alarm system (exclusive of any connected to a public emergency reporting system) in which alarm, supervisory, or trouble signals are transmitted automatically to, recorded in, and supervised from a remote supervising station that has competent and experienced servers and operators who, upon receipt of a signal, take such action as required by this Code. (SIG-SSS)

3.3.282 Supervising Station Service.

3.3.282.1 ***Central Station Service.*** The use of a system or a group of systems including the protected premises fire alarm system(s) in which the operations of circuits and devices are signaled to, recorded in, and supervised from a listed central station that has competent and experienced operators who, upon receipt of a signal, take such action as required by this Code. Related activities at the protected premises, such as equipment installation, inspection, testing, maintenance, and runner service, are the responsibility of the central station or a listed alarm service local company. Central station service is controlled and operated by a person, firm, or corporation whose business is the furnishing of such contracted services or whose properties are the protected premises. (SIG-SSS)

3.3.282.2 ***Proprietary Supervising Station Service.*** The use of a system or a group of systems including the protected premises fire alarm system(s) in which the operations of circuits and devices are signaled to, recorded in, and supervised from a supervising station under the same ownership as the protected premises that has competent and experienced operators who, upon receipt of a signal, take such action as required by this Code. Related activities at the protected premises, such as equipment installation, inspection, testing, maintenance, and runner service, are the responsibility of the owner. Proprietary supervising station service is controlled and operated by the entity whose properties are the protected premises. (SIG-SSS)

3.3.282.3 ***Remote Supervising Station Service.*** The use of a system including the protected premises fire alarm system(s) in which the operations of circuits and devices are signaled to, recorded in, and supervised from a supervising station that has competent and experienced operators who, upon receipt of a signal, take such action as required by this Code. Related activities at the protected premises, such as equipment installation, inspection, testing, and maintenance, are the responsibility of the owner. (SIG-SSS)

3.3.283 Supervisory Service. The service required to monitor performance of guard tours and the operative condition of fixed suppression systems or other systems for the protection of life and property. (SIG-PRO)

3.3.284 Supervisory Signal. See 3.3.253, Signal.

3.3.285 Supervisory Signal Initiating Device. See 3.3.131, Initiating Device.

3.3.286 Supplementary. As used in this Code, *supplementary* refers to equipment or operations not required by this Code and designated as such by the authority having jurisdiction. (SIG-FUN)

3.3.287 Switched Telephone Network.

3.3.287.1 ***Loop Start Telephone Circuit.*** A loop start telephone circuit is an analog telephone circuit that supports loop start signaling as specified in either Telcordia *GR-506-CORE, LATA Switching Systems Generic Requirements: Signaling for Analog Interface*, or Telcordia *GR-909-CORE, Fiber in the Loop Systems Generic Requirements.* (SIG-SSS)

3.3.287.2 ***Public Switched Telephone Network.*** An assembly of communications equipment and telephone service providers that utilize managed facilities-based voice networks (MFVN) to provide the general public with the ability to establish communications channels via discrete dialing codes. (SIG-SSS)

3.3.288 System Operator. An individual trained to operate and/or initiate a mass notification system. (SIG-ECS)

3.3.289 System Unit. The active subassemblies at the supervising station used for signal receiving, processing, display, or recording of status change signals; a failure of one of these subassemblies causes the loss of a number of alarm signals by that unit. (SIG-SSS)

3.3.290 Tactile Notification Appliance. See 3.3.172, Notification Appliance.

3.3.291 Talk Mode. A means of communications within a building normally dedicated to emergency functions. Commonly referred to as fire fighters' phones, but can also be used for communications with fire fighters and/or fire wardens, including occupants, during an emergency, such as between a fire command center and a designated location, such as a stair, stairwell, or location of emergency equipment. (SIG-ECS)

3.3.291.1 ***Common Talk Mode.*** The ability to conference multiple telephones in a single conversation. This is similar to what was referred to as a party line. (SIG-ECS)

3.3.291.2 ***Selective Talk Mode.*** The ability for personnel at the fire command center to receive indication of incoming calls and choose which call to answer. This includes the ability to transfer between incoming calls and conference multiple phone locations. Selective calling can include the ability to initiate calls to emergency phone locations. (SIG-ECS)

3.3.292 Testing Personnel. See 3.3.190, Personnel.

3.3.293 Textual Audible Notification Appliance. See 3.3.172, Notification Appliance.

3.3.294 Textual Visible Notification Appliance. See 3.3.172, Notification Appliance.

3.3.295 Transmission Channel. See 3.3.44, Channel.

3.3.296 Transmitter. A system component that provides an interface between signaling line circuits, initiating device circuits, or control units and the transmission channel. (SIG-SSS)

3.3.297 Transponder. A multiplex alarm transmission system functional assembly located at the protected premises. (SIG-SSS)

3.3.298 Trouble Signal. See 3.3.253, Signal.

3.3.299 Two-Way Emergency Communications System. See 3.3.85, Emergency Communications System.

3.3.300 Type A Public Emergency Alarm Reporting System. See 3.3.211, Public Emergency Alarm Reporting System.

3.3.301 Type B Public Emergency Alarm Reporting System. See 3.3.211, Public Emergency Alarm Reporting System.

3.3.302 Unintentional Alarm. See 3.3.304.3.

3.3.303 Unknown Alarm. See 3.3.304.4.

3.3.304* Unwanted Alarm. Any alarm that occurs that is not the result of a potentially hazardous condition. (SIG-FUN)

3.3.304.1 ***Malicious Alarm.*** An unwanted activation of an alarm initiating device caused by a person acting with malice. (SIG-FUN)

3.3.304.2* ***Nuisance Alarm.*** An unwanted activation of a signaling system or an alarm initiating device in response to a stimulus or condition that is not the result of a potentially hazardous condition. (SIG-FUN)

3.3.304.3 ***Unintentional Alarm.*** An unwanted activation of an alarm initiating device caused by a person acting without malice. (SIG-FUN)

3.3.304.4 ***Unknown Alarm.*** An unwanted activation of an alarm initiating device or system output function where the cause has not been identified. (SIG-FUN)

3.3.305 Uplink. The radio signal from the portable public safety subscriber transmitter to the base station receiver. (SIG-ECS)

3.3.306* Video Image Flame Detection (VIFD). The principle of using automatic analysis of real-time video images to detect the presence of flame. (SIG-IDS)

3.3.307 Video Image Smoke Detection (VISD). See 3.3.266, Smoke Detection.

3.3.308 Visible Notification Appliance. See 3.3.172, Notification Appliance.

3.3.309 Voice Message Priority. A scheme for prioritizing mass notification messages. (SIG-ECS)

3.3.310 WATS (Wide Area Telephone Service). Telephone company service allowing reduced costs for certain telephone call arrangements. In-WATS or 800-number service calls can be placed from anywhere in the continental United States to the called party at no cost to the calling party. Out-WATS is a service whereby, for a flat-rate charge, dependent on the total duration of all such calls, a subscriber can make an unlimited number of calls within a prescribed area from a particular telephone terminal without the registration of individual call charges. (SIG-SSS)

3.3.311* Wavelength. The distance between the peaks of a sinusoidal wave. All radiant energy can be described as a wave having a wavelength. Wavelength serves as the unit of measure for distinguishing between different parts of the spectrum. Wavelengths are measured in microns (μm), nanometers (nm), or angstroms (Å). (SIG-IDS)

3.3.312 Wide-Area Mass Notification System. See 3.3.85, Emergency Communications System.

3.3.313 Wide-Area Signaling. Signaling intended to provide alerting or information to exterior open spaces, such as campuses, neighborhood streets, a city, a town, or a community. (SIG-NAS)

3.3.314 Wireless Control Unit. See 3.3.60, Control Unit.

3.3.315 Wireless Protection System. A system or a part of a system that can transmit and receive signals without the aid of interconnection wiring. It can consist of either a wireless control unit or a wireless repeater. (SIG-PRO)

3.3.316 Wireless Repeater. A component used to relay signals among wireless devices, appliances, and control units. (SIG-PRO)

3.3.317 Zone. A defined area within the protected premises. A zone can define an area from which a signal can be received, an area to which a signal can be sent, or an area in which a form of control can be executed. (SIG-FUN)

3.3.317.1 ***Notification Zone.*** A discrete area of a building, bounded by building outer walls, fire or smoke compartment boundaries, floor separations, or other fire safety subdivisions, in which occupants are intended to receive common notification. (SIG-PRO)

3.3.317.2* ***Signaling Zone.*** An area consisting of one or more notification zones where signals are actuated simultaneously. (SIG-ECS)

Chapter 4 Reserved

Chapter 5 Reserved

Chapter 6 Reserved

Chapter 7 Documentation

7.1 Application. (SIG-FUN)

7.1.1 The documentation of the design, acceptance, and completion of new systems required under this Code shall comply with the minimum requirements of this chapter.

7.1.2 The documentation of the alteration, maintenance, and testing of existing systems shall comply with the minimum requirements of this chapter.

7.1.3* Where required by governing laws, codes, or standards, or other parts of this Code, the requirements of this chapter, or portions thereof, shall apply.

7.1.4 Unless required by other governing laws, codes, or standards, the documentation requirements of this chapter shall not apply to Chapter 29.

7.1.5 This chapter outlines documentation requirements but does not prohibit additional documentation from being provided.

7.1.6 The requirements of Chapters 10, 12, 14, 17, 18, 21, 23, 24, 26, and 27 shall apply unless otherwise noted in this chapter.

7.2* Minimum Required Documentation. (SIG-FUN)

7.2.1* Where documentation is required by the authority having jurisdiction, the following list shall represent the minimum documentation required for new fire alarm systems, supervising station and shared communication equipment, and emergency communications systems, including new systems and additions or alterations to existing systems:

(1)*Written narrative providing intent and system description
(2) Riser diagram
(3) Floor plan layout showing locations of all devices, control equipment, and supervising station and shared communications equipment with each sheet showing the following:
 (a) Point of compass (north arrow)
 (b) A graphic representation of the scale used
 (c) Room use identification
 (d) Building features that will affect the placement of initiating devices and notification appliances
(4) Sequence of operation in either an input/output matrix or narrative form
(5) Equipment technical data sheets
(6) Manufacturers' published instructions, including operation and maintenance instructions
(7) Battery capacity and de-rating calculations (where batteries are provided)
(8) Voltage drop calculations for notification appliance circuits
(9) Mounting height elevation for wall-mounted devices and appliances
(10) Where occupant notification is required, minimum sound pressure levels that must be produced by the audible notification appliances in applicable covered areas
(11) Pathway diagrams between the control unit and the supervising station and shared communications equipment
(12) Completed record of completion in accordance with 7.5.6 and 7.8.2
(13) For software-based systems, a copy of site-specific software, including specific instructions on how to obtain the means of system and software access (password)
(14) Record (as-built) drawings
(15) Records, record retention, and record maintenance in accordance with Section 7.7
(16) Completed record of inspection and testing in accordance with 7.6.6 and 7.8.2

7.2.2 System design documents shall identify the name and contact information of the system designer.

7.2.3 All fire alarm drawings shall use symbols described in NFPA 170 or other symbols acceptable to the authority having jurisdiction.

7.3 Design (Layout) Documentation.

7.3.1* The requirements of Section 7.3 shall apply only where required by other governing laws, codes, or standards; by other parts of this Code; or by project specifications or drawings. (SIG-FUN)

7.3.2* Where required by governing laws, codes, or standards, or other parts of this Code, design (layout) documents shall be prepared prior to installing new systems. (SIG-ECS)

7.3.3* Where required by governing laws, codes, or standards, or other parts of this Code, preliminary plans shall be created. (SIG-ECS)

7.3.4 Notification. (SIG-NAS)

7.3.4.1* The requirements of 7.3.4 shall apply only where required by other governing laws, codes, or standards, or by other parts of this Code.

7.3.4.2 Design documents shall include ambient sound pressure levels and audible design sound pressure levels in accordance with 18.4.1.4.3.

7.3.4.3 Analysis and design documentation for narrow band tone signaling shall be in accordance with 18.4.6.4.

7.3.4.4 The documentation of acoustically distinguishable spaces (ADS) shall be in accordance with 18.4.10.

7.3.4.5 Design documents shall specify the rooms and spaces that will have visible notification and those where visible notification will not be provided in accordance with 18.5.2.1.

7.3.4.6 Performance-based design alternatives for strobe design shall be in accordance with 18.5.5.6.2.

7.3.5 Detection. (SIG-IDS)

7.3.5.1 Heat-Sensing Fire Detectors. Heat detection design documentation shall be provided in accordance with Section 17.6.

7.3.5.2 Smoke-Sensing Fire Detectors. Smoke detection design documentation shall be provided in accordance with Section 17.7.

7.3.5.3 Radiant Energy-Sensing Fire Detectors. Radiant energy detection design documentation shall be provided in accordance with Section 17.8.

7.3.5.4 Gas Detectors. Gas detection design documentation shall be provided in accordance with Section 17.10.

7.3.6* Risk Analysis Documentation. (SIG-ECS)

7.3.6.1 When a risk analysis is required to be prepared, findings and considerations of the risk analysis shall be documented.

7.3.6.2 When determined by the stakeholders, security and protection of the risk analysis documentation shall be in accordance with 7.3.7 and Section 7.7.

7.3.6.3 The risk analysis documentation shall list the various scenarios evaluated and the anticipated outcomes.

7.3.6.4 Risk analyses for mass notification systems shall be documented in accordance with 7.3.6 and 24.3.11.

7.3.7* Performance-Based Design Documentation.

7.3.7.1 Performance-based design documentation for fire detection shall be in accordance with Section 17.3. (SIG-IDS)

7.3.7.2 Performance-based design documentation for strobes shall be in accordance with 18.5.5.6.2. (SIG-NAS)

7.3.7.3 A copy of approval documentation resulting from performance-based designs shall be included with the completion documentation in accordance with 7.5. (SIG-FUN)

7.3.7.4 Performance-based design documentation for signaling line circuit zoning shall be in accordance with 23.6.1.4 and 23.6.1.5. (SIG-PRO)

7.3.8 Emergency Response Plan Documentation. (SIG-ECS)

7.3.8.1 When an emergency response plan is required to be prepared, such as for a mass notification system, findings of the plan shall be documented.

7.3.8.2 When identified by the stakeholders, security and protection of the emergency response plan documentation shall be in accordance with 7.7.3.

7.3.8.3 The emergency response plan shall document the various scenarios evaluated and the anticipated outcomes.

7.3.9 Evaluation Documentation. (SIG-FUN)

7.3.9.1* Evaluation documentation shall include a signed statement(s) by the person responsible for the design attesting to the evaluation and the resultant technical decision and deeming it reliable and acceptable for the particular application.

7.3.9.2 A copy of the evaluation documentation shall be retained for the life of the system and be maintained with the documents required by 7.7.1.6.

7.4 Shop Drawings (Installation Documentation). (SIG-FUN)

7.4.1* The requirements of Section 7.4 shall apply only where required by other governing laws, codes, or standards; by other parts of this Code; or by project specifications or drawings.

7.4.2* Shop drawings shall be drawn to an indicated scale, on sheets of uniform size, with a plan of each floor.

7.4.3 Shop drawings for fire alarm and emergency communications systems shall provide basic information and shall provide the basis for the record (as-built) drawings required in accordance with 7.5.2.

7.4.4 Shop drawings shall include the following information:

(1) Name of protected premises, owner, and occupant (where applicable)
(2) Name of installer or contractor
(3) Location of protected premises
(4) Device legend and symbols in accordance with NFPA 170, or other symbols acceptable to the authority having jurisdiction
(5) Date of issue and any revision dates

7.4.5 Floor plan drawings shall be drawn to an indicated scale and shall include the following information, where applicable for the particular system:

(1) Floor or level identification
(2) Point of compass (indication of North)
(3) Graphic scale
(4) All walls and doors

(5) All partitions extending to within 15 percent of the ceiling height (where applicable and when known)
(6) Room and area descriptions
(7) System devices/component locations
(8) Locations of fire alarm primary power disconnecting means
(9) Locations of monitor/control interfaces to other systems
(10) System riser locations
(11) Type and number of system components/devices on each circuit, on each floor or level
(12) Type and quantity of conductors and conduit (if used) for each circuit
(13) Identification of any ceiling over 10 ft (3.0 m) in height where automatic fire detection is being proposed
(14) Details of ceiling geometries, including beams and solid joists, where automatic fire detection is being proposed
(15) Where known, acoustic properties of spaces

7.4.6 System riser diagrams shall be coordinated with the floor plans and shall include the following information:

(1) General arrangement of the system in building cross-section
(2) Number of risers
(3) Type and number of circuits in each riser
(4) Type and number of system components/devices on each circuit, on each floor or level
(5) Number of conductors for each circuit

7.4.7 Control unit diagrams shall be provided for all control equipment (i.e., equipment listed as either a control unit or control unit accessory), power supplies, battery chargers, and annunciators and shall include the following information:

(1) Identification of the control equipment depicted
(2) Location(s) of control equipment
(3) All field wiring terminals and terminal identifications
(4) All circuits connected to field wiring terminals and circuit identifications
(5) All indicators and manual controls
(6) Field connections to supervising station signaling equipment, releasing equipment, or emergency safety control interfaces, where provided

7.4.8 Typical wiring diagrams shall be provided for all initiating devices, notification appliances, remote indicators, annunciators, remote test stations, and end-of-line and power supervisory devices.

7.4.9* A narrative description or input/output matrix of operation shall be provided to describe the sequence of operation.

7.4.10 System calculations shall be included as follows:

(1) Battery calculations
(2) Notification appliance circuit voltage drop calculations
(3) Other required calculations, such as line resistance calculations, where required

7.5 Completion Documentation.

7.5.1* The requirements of Section 7.5 shall apply only where required by other governing laws, codes, or standards; by other parts of this Code; or by project specifications or drawings. (SIG-FUN)

7.5.2 Before requesting final approval of the installation, if required by the authority having jurisdiction, the installing contractor shall furnish a written statement stating that the system has been installed in accordance with approved plans and tested in accordance with the manufacturer's published instructions and the appropriate NFPA requirements. (SIG-FUN)

7.5.3 All systems including new systems and additions or alterations to existing systems shall include the following documentation, which shall be delivered to the owner or the owner's representative upon final acceptance of the system:

(1)*An owner's manual and manufacturer's published instructions covering all system equipment
(2) Record (as-built) drawings in accordance with 7.5.5
(3) A completed record of completion form in accordance with 7.5.6
(4) For software-based systems, record copy of the site-specific software in accordance with 7.5.7 (SIG-FUN)

7.5.4 For new emergency communications systems, an owner's manual shall be provided and shall contain the following documentation:

(1) Detailed narrative description of the system inputs, evacuation signaling, ancillary functions, annunciation, intended sequence of operations, expansion capability, application considerations, and limitations
(2) Written sequence of operation for the system including an operational input/output matrix
(3) Operator instructions for basic system operations, including alarm acknowledgment, system reset, interpretation of system output (LEDs, CRT display, and printout), operation of manual evacuation signaling and ancillary function controls, and change of printer paper
(4) Detailed description of routine maintenance and testing as required and recommended and as would be provided under a maintenance contract, including testing and maintenance instructions for each type of device installed, which includes the following:
    (a) Listing of the individual system components that require periodic testing and maintenance
    (b) Step-by-step instructions detailing the requisite testing and maintenance procedures, and the intervals at which those procedures shall be performed, for each type of device installed
    (c) Schedule that correlates the testing and maintenance procedures that are required by this section
(5) Service directory, including a list of names and telephone numbers of those who provide service for the system
(6) Product data sheets for all system equipment (SIG-ECS)

7.5.5 Record Drawings (As-Builts). (SIG-FUN)

7.5.5.1 Record drawings shall consist of current updated shop drawings reflecting the actual installation of all system equipment, components, and wiring.

7.5.5.2* A sequence of operations in input/output matrix or narrative form shall be provided with the record drawings to reflect actual programming at the time of completion.

7.5.5.3 Where necessary, revised calculations in accordance with 7.4.10 shall be provided depicting any changes due to installation conditions.

7.5.5.4 Record drawings shall be turned over to the owner with a copy placed inside the documentation cabinet in accordance with Section 7.7.

7.5.5.5* Record drawings shall include approval documentation resulting from variances, performance-based designs, risk analyses, and other system evaluations or variations.

7.5.6 Record of Completion. (SIG-FUN)

7.5.6.1* The record of completion shall be documented in accordance with 7.5.6 using either the record of completion forms, Figure 7.8.2(a) through Figure 7.8.2(f), or an alternative document that contains only the elements of Figure 7.8.2(a) through Figure 7.8.2(f) applicable to the installed system.

7.5.6.2* The record of completion documentation shall be completed by the installing contractor and submitted to the enforcing authority and the owner at the conclusion of the job. The record of completion documentation shall be permitted to be part of the written statement required in 7.5.2 and part of the documents that support the requirements of 7.5.8. When more than one contractor has been responsible for the installation, each contractor shall complete the portions of the documentation for which that contractor has responsibility.

7.5.6.3* The preparation of the record of completion documentation shall be the responsibility of the qualified and experienced person in accordance with 10.5.2.

7.5.6.4 The record of completion documentation shall be updated to reflect all system additions or modifications.

7.5.6.5 The updated copy of the record of completion documents shall be maintained in a documentation cabinet in accordance with 7.7.2.

7.5.6.6 Revisions.

7.5.6.6.1 All modifications made after the initial installation shall be recorded on a revised version of the original completion documents.

7.5.6.6.2 The revised record of completion document shall include a revision date.

7.5.6.6.3* Where the original or the latest overall system record of completion cannot be obtained, a new system record of completion shall be provided that documents the system configuration as discovered during the current project's scope of work.

7.5.6.7 Electronic Record of Completion.

7.5.6.7.1 Where approved by the authority having jurisdiction, the record of completion shall be permitted to be filed electronically instead of on paper.

7.5.6.7.2 If filed electronically, the record of completion document shall be accessible with standard software and shall be backed up.

7.5.7 Site-Specific Software. (SIG-TMS)

7.5.7.1 For software-based systems, a copy of the site-specific software shall be provided to the system owner or owner's designated representative.

7.5.7.1.1 The site-specific software documentation shall include both the user passcode and either the system programming password or specific instructions on how to obtain the programming password from the system manufacturer.

7.5.7.1.2 The passwords provided shall enable currently certified qualified programming personnel to access, edit, modify, and add to the existing system site-specific software.

7.5.7.2 A copy of the site-specific software shall be stored onsite in nonvolatile, nonerasable, nonrewritable memory.

7.5.8* Verification of Compliant Installation. (SIG-FUN)

7.5.8.1 Where required by the authority having jurisdiction, compliance of the completed installation with the requirements of this Code shall be certified by a qualified and impartial third-party organization acceptable to the authority having jurisdiction.

7.5.8.2 Verification of compliant installation shall be performed according to testing requirements and procedures specified in 14.4.1 and 14.4.2.

7.5.8.3 Verification shall ensure that:

(1) All components and functions are installed and operate per the approved plans and sequence of operation.
(2) All required system documentation is complete and is archived on site.
(3) For new supervising station systems, the verification shall also ascertain proper arrangement, transmission, and receipt of all signals required to be transmitted off-premises and shall meet the requirements of 14.4.1 and 14.4.2.
(4) For existing supervising station systems that are extended, modified, or reconfigured, the verification shall be required for the new work only, and reacceptance testing in accordance with Chapter 14 shall be acceptable.
(5) Written confirmation has been provided that any required corrective actions have been completed

7.5.9 Documentation of central station service shall be in accordance with 26.3.4. (SIG-SSS)

7.5.10 Documentation of remote station service shall be in accordance with 26.5.2. (SIG-SSS)

7.6 Inspection, Testing, and Maintenance Documentation. (SIG-TMS)

7.6.1 Test plan documentation shall be provided in accordance with 14.2.10.

7.6.2 Acceptance testing documentation shall be provided in accordance with 14.6.1.

7.6.3 Reacceptance test documentation shall be provided in accordance with 14.6.1.

7.6.4 Periodic inspection and testing documentation shall be provided in accordance with 14.6.2 through 14.6.4.

7.6.5 Impairment documentation shall be provided in accordance with Section 10.20.

7.6.6 Record of Inspection and Testing. The record of all inspections, testing, and maintenance as required by 14.6.2.4 shall be documented using either the record of inspection and testing forms, Figure 7.8.2(g) through Figure 7.8.2(l), or an alternative record that includes all the applicable information shown in Figure 7.8.2(g) through Figure 7.8.2(l).

7.7 Records, Record Retention, and Record Maintenance.

7.7.1 Records. (SIG-FUN)

7.7.1.1 A complete record of the tests and operations of each system shall be kept until the next test and for 1 year thereafter unless more stringent requirements are required elsewhere in this Code.

7.7.1.2* The records shall be available for examination and, if required, reported to the authority having jurisdiction. Archiving of records by any means shall be permitted if hard copies of the records can be provided promptly when requested.

7.7.1.3 If off-premises monitoring is provided, records of all signals, tests, and operations recorded at the supervising station, including the public emergency alarm reporting system, shall be maintained by the off-premise monitoring service provider for not less than 1 year unless more stringent requirements are required elsewhere in this Code.

7.7.1.4 Required documents regarding system design and function shall be maintained for the life of the system.

7.7.1.5 The emergency communications system and fire alarm system as-built plans and other related documentation shall be permitted to be maintained together, including the appearance of both systems on the same drawings. (SIG-ECS)

7.7.1.6 Revisions and alterations to systems shall be recorded and records maintained with the original system design documents.

7.7.2 Document Accessibility. (SIG-FUN)

7.7.2.1 With every new system, a documentation cabinet shall be installed at the system control unit or at another approved location at the protected premises.

7.7.2.2 The documentation cabinet shall be sized so that it can contain all necessary documentation.

7.7.2.3* All record documentation shall be stored in the documentation cabinet.

7.7.2.4 Where the documentation cabinet is not in the same location as the system control unit, its location shall be identified at the system control unit.

7.7.2.5 The documentation cabinet shall be prominently labeled SYSTEM RECORD DOCUMENTS.

7.7.2.6* The building owner or the building owner's representative shall, on an annual basis, review any electronic documentation media formats and associated interfacing hardware for compatibility and update, if necessary.

7.7.2.7 The contents of the cabinet shall be accessible by authorized personnel only.

7.7.2.8 Emergency communications system and fire alarm system record documentation shall be permitted to be maintained together in the same documentation cabinet.

7.7.3 Document Security. (SIG-ECS)

7.7.3.1 Security for system's documentation shall be determined by the stakeholders.

7.7.3.2* Where such documents cannot be protected from public access, it shall be permitted to remove sensitive information from record documents provided the owner retains complete documentation that will be made accessible to the authority having jurisdiction at an owner designated location.

7.8 Forms.

7.8.1 General.

7.8.1.1* The requirements of Section 7.8 shall apply only where required by other governing laws, codes, or standards; by other parts of this Code; or by project specifications or drawings. (SIG-FUN)

7.8.1.2 Where specific forms are required by other governing laws, codes, or standards; by other parts of this Code; or by project specifications or drawings, form layouts and content that differ from those in Section 7.8 shall be permitted provided that the minimum required content is included. (SIG-FUN)

7.8.2 Forms for Documentation. Forms for documentation shall be as follows:

(1)*Unless otherwise permitted or required in 7.5.6 or 7.8.1.2, Figure 7.8.2(a) through Figure 7.8.2(f) shall be used to document the record of completion. (SIG-FUN)

(2)*Unless otherwise permitted or required in 7.6.6 or 7.8.1.2, Figure 7.8.2(g) through Figure 7.8.2(l) shall be used to document the record of inspection and testing. (SIG-TMS)

SYSTEM RECORD OF COMPLETION

This form is to be completed by the system installation contractor at the time of system acceptance and approval.
It shall be permitted to modify this form as needed to provide a more complete and/or clear record.
Insert N/A in all unused lines.
Attach additional sheets, data, or calculations as necessary to provide a complete record.

Form Completion Date: ________________ Supplemental Pages Attached: __________

1. PROPERTY INFORMATION

Name of property: ____________________

Address: ____________________

Description of property: ____________________

Name of property representative: ____________________

Address: ____________________

Phone: ____________ Fax: ____________ E-mail: ____________

2. INSTALLATION, SERVICE, TESTING, AND MONITORING INFORMATION

Installation contractor: ____________________

Address: ____________________

Phone: ____________ Fax: ____________ E-mail: ____________

Service organization: ____________________

Address: ____________________

Phone: ____________ Fax: ____________ E-mail: ____________

Testing organization: ____________________

Address: ____________________

Phone: ____________ Fax: ____________ E-mail: ____________

Effective date for test and inspection contract: ____________________

Monitoring organization: ____________________

Address: ____________________

Phone: ____________ Fax: ____________ E-mail: ____________

Account number: ____________ Phone line 1: ____________ Phone line 2: ____________

Means of transmission: ____________________

Entity to which alarms are retransmitted: ____________ Phone: ____________

3. DOCUMENTATION

On-site location of the required record documents and site-specific software: ____________________

4. DESCRIPTION OF SYSTEM OR SERVICE

This is a: ❑ New system ❑ Modification to existing system Permit number: ____________

NFPA 72 edition: __________

4.1 Control Unit

Manufacturer: ____________________ Model number: ____________

4.2 Software and Firmware

Firmware revision number: ____________________

4.3 Alarm Verification

❑ This system does not incorporate alarm verification.

Number of devices subject to alarm verification: ____________ Alarm verification set for ________ seconds

 NFPA 72 (p. 1 of 3)

FIGURE 7.8.2(a) System Record of Completion. (SIG-FUN)

SYSTEM RECORD OF COMPLETION *(continued)*

5. SYSTEM POWER

5.1 Control Unit

5.1.1 Primary Power

Input voltage of control panel: ____________________ Control panel amps: ____________________

Overcurrent protection: Type: ____________________ Amps: ____________________

Branch circuit disconnecting means location: ____________________ Number: ____________________

5.1.2 Secondary Power

Type of secondary power: ____________________

Location, if remote from the plant: ____________________

Calculated capacity of secondary power to drive the system:

In standby mode (hours): ____________________ In alarm mode (minutes): ____________________

5.2 Control Unit

❑ This system does not have power extender panels

❑ Power extender panels are listed on supplementary sheet A

6. CIRCUITS AND PATHWAYS

Pathway Type	Dual Media Pathway	Separate Pathway	Class	Survivability Level
Signaling Line				
Device Power				
Initiating Device				
Notification Appliance				
Other (specify):				

7. REMOTE ANNUNCIATORS

Type	Location

8. INITIATING DEVICES

Type	Quantity	Addressable or Conventional	Alarm or Supervisory	Sensing Technology
Manual Pull Stations				
Smoke Detectors				
Duct Smoke Detectors				
Heat Detectors				
Gas Detectors				
Waterflow Switches				
Tamper Switches				

 NFPA 72 (p. 2 of 3)

FIGURE 7.8.2(a) ***Continued***

SYSTEM RECORD OF COMPLETION *(continued)*

9. NOTIFICATION APPLIANCES

Type	Quantity	Description
Audible		
Visible		
Combination Audible and Visible		

10. SYSTEM CONTROL FUNCTIONS

Type	Quantity
Hold-Open Door Releasing Devices	
HVAC Shutdown	
Fire/Smoke Dampers	
Door Unlocking	
Elevator Recall	
Elevator Shunt Trip	

11. INTERCONNECTED SYSTEMS

❑ This system does not have interconnected systems.

❑ Interconnected systems are listed on supplementary sheet ________.

12. CERTIFICATION AND APPROVALS

12.1 System Installation Contractor

This system as specified herein has been installed according to all NFPA standards cited herein.

Signed: ____________________ Printed name: ____________________ Date: ____________

Organization: ____________________ Title: ____________________ Phone: ____________

12.2 System Operational Test

This system as specified herein has tested according to all NFPA standards cited herein.

Signed: ____________________ Printed name: ____________________ Date: ____________

Organization: ____________________ Title: ____________________ Phone: ____________

12.3 Acceptance Test

Date and time of acceptance test: ____________________

Installing contractor representative: ____________________

Testing contractor representative: ____________________

Property representative: ____________________

AHJ representative: ____________________

 NFPA 72 (p. 3 of 3)

FIGURE 7.8.2(a) ***Continued***

EMERGENCY COMMUNICATIONS SYSTEMS SUPPLEMENTARY RECORD OF COMPLETION

This form is a supplement to the System Record of Completion. It includes systems and components specific to emergency communications systems.
This form is to be completed by the system installation contractor at the time of system acceptance and approval.
It shall be permitted to modify this form as needed to provide a more complete and/or clear record.
Insert N/A in all unused lines.

Form Completion Date: ________________ Number of Supplemental Pages Attached: __________

1. PROPERTY INFORMATION

Name of property: ______________________________

Address: ______________________________

2. DESCRIPTION OF SYSTEM OR SERVICE

❑ Fire alarm with in-building fire emergency voice alarm communication system (EVAC)

❑ Mass notification system

❑ Combination system, with the following components:

❑ Fire alarm ❑ EVACS ❑ MNS ❑ Two-way, in-building, emergency communications system

❑ Other (specify): ______________________________

NFPA 72 edition: __________ Additional description of system(s): ______________________________

____

2.1 In-Building Fire Emergency Voice Alarm Communications System

Manufacturer: ______________________________ Model number: ________________

Number of single voice alarm channels: ____________ Number of multiple voice alarm channels: ____________

Number of speakers: ____________ Number of speaker circuits: ____________

Location of amplification and sound processing equipment: ______________________________

____

____

Location of paging microphone stations:

Location 1: ______________________________

Location 2: ______________________________

Location 3: ______________________________

2.2 Mass Notification System

2.2.1 System Type:

❑ In-building MNS–combination

❑ In-building MNS ❑ Wide-area MNS ❑ Distributed recipient MNS

❑ Other (specify): ______________________________

 NFPA 72 (p. 1 of 3)

FIGURE 7.8.2(b) Emergency Communications System Supplementary Record of Completion. (SIG-FUN)

EMERGENCY COMMUNICATIONS SYSTEMS SUPPLEMENTARY RECORD OF COMPLETION *(continued)*

2. DESCRIPTION OF SYSTEM OR SERVICE *(continued)*

2.2.2 System Features:

❑ Combination fire alarm/MNS ❑ MNS autonomous control unit ❑ Wide-area MNS to regional national alerting interface

❑ Local operating console (LOC) ❑ Distributed-recipient MNS (DRMNS) ❑ Wide-area MNS to DRMNS interface

❑ Wide-area MNS to high power speaker array (HPSA) interface ❑ In-building MNS to wide-area MNS interface

❑ Other (specify): ____________________

2.2.3 MNS Local Operating Consoles

Location 1: ____________________

Location 2: ____________________

Location 3: ____________________

2.2.4 High Power Speaker Arrays

Number of HPSA speaker initiation zones: ____________________

Location 1: ____________________

Location 2: ____________________

Location 3: ____________________

2.2.5 Mass Notification Devices

Combination fire alarm/MNS visual devices: __________ MNS-only visual devices: __________

Textual signs: __________ Other (describe): ____________________

Supervision class: ____________________

2.2.6 Special Hazard Notification

❑ This system does not have special suppression predischarge notification.

❑ MNS systems DO NOT override notification appliances required to provide special suppression predischarge notification.

3. TWO-WAY EMERGENCY COMMUNICATIONS SYSTEMS

3.1 Telephone System

Number of telephone jacks installed: __________ Number of warden stations installed: __________

Number of telephone handsets stored on site: ____________________

Type of telephone system installed: ❑ Electrically powered ❑ Sound powered

3.2 Area of Refuge (Area of Rescue Assistance) Emergency Communications Systems

Number of stations: __________ Location of central control point: ____________________

Days and hours when central control point is attended: ____________________

Location of alternate control point: ____________________

Days and hours when alternate control point is attended: ____________________

 NFPA 72 (p. 2 of 3)

FIGURE 7.8.2(b) ***Continued***

EMERGENCY COMMUNICATIONS SYSTEMS SUPPLEMENTARY RECORD OF COMPLETION *(continued)*

3. TWO-WAY EMERGENCY COMMUNICATIONS SYSTEMS *(continued)*

3.3 Elevator Emergency Communications Systems

Number of elevators with stations: ________ Location of central control point: ________

Days and hours when central control point is attended: ________

Location of alternate control point: ________

Days and hours when alternate control point is attended: ________

3.4 Other Two-Way Communications System

Describe: ________

4. CONTROL FUNCTIONS

This system activates the following control functions specific to emergency communications systems:

Type	Quantity
Mass Notification Override of Alarm Signaling Systems or Appliances	

See Main System Record of Completion for additional information, certifications, and approvals.

 NFPA 72 (p. 3 of 3)

FIGURE 7.8.2(b) ***Continued***

POWER SYSTEMS
SUPPLEMENTARY RECORD OF COMPLETION

This form is a supplement to the System Record of Completion. It includes systems and components specific to power systems that incorporate generators, UPS systems, remote battery systems, or other complex power systems. This form is to be completed by the system installation contractor at the time of system acceptance and approval. It shall be permitted to modify this form as needed to provide a more complete and/or clear record. Insert N/A in all unused lines.

Form Completion Date: __________ Number of Supplemental Pages Attached: __________

1. PROPERTY INFORMATION

Name of property: __________

Address: __________

2. SYSTEM POWER

2.1 Control Unit

2.1.1 Primary Power

Input voltage of control panel: __________ Control panel amps: __________

Overcurrent protection: Type: __________ Amps: __________

Location (of primary supply panelboard): __________

Disconnecting means location: __________

2.1.2 Engine-Driven Generator

Location of generator: __________

Location of fuel storage: __________ Type of fuel: __________

2.1.3 Uninterruptible Power System

Equipment powered by UPS system: __________

Location of UPS system: __________

Calculated capacity of UPS batteries to drive the system components connected to it:

In standby mode (hours): __________ In alarm mode (minutes): __________

2.1.4 Batteries

Location: __________ Type: __________ Nominal voltage: __________ Amp/hour rating: __________

Calculated capacity of batteries to drive the system:

In standby mode (hours): __________ In alarm mode (minutes): __________

2.2 In-Building Fire Emergency Voice Alarm Communications System or Mass Notification System

2.2.1 Primary Power

Input voltage of EVACS or MNS panel: __________ EVACS or MNS panel amps: __________

Overcurrent protection: Type: __________ Amps: __________

Location (of primary supply panelboard): __________

Disconnecting means location: __________

 NFPA 72 (p. 1 of 3)

FIGURE 7.8.2(c) Power Systems Supplementary Record of Completion. (SIG-FUN)

POWER SYSTEMS
SUPPLEMENTARY RECORD OF COMPLETION *(continued)*

2. SYSTEM POWER *(continued)*

2.2.2 Engine-Driven Generator

Location of generator: ____________________

Location of fuel storage: ____________________ Type of fuel: ____________________

2.2.3 Uninterruptible Power System

Equipment powered by UPS system: ____________________

Location of UPS system: ____________________

Calculated capacity of UPS batteries to drive the system components connected to it:

In standby mode (hours): ____________________ In alarm mode (minutes): ____________________

2.2.4 Batteries

Location: ____________ Type: ________ Nominal voltage: ________ Amp/hour rating: ________

Calculated capacity of batteries to drive the system:

In standby mode (hours): ____________________ In alarm mode (minutes): ____________________

2.3 Notification Appliance Power Extender Panels

❑ This system does not have power extender panels.

2.3.1 Primary Power

Input voltage of power extender panel(s): ____________ Power extender panel amps: ____________

Overcurrent protection: Type: ____________ Amps: ____________

Location (of primary supply panelboard): ____________________

Disconnecting means location: ____________________

2.3.2 Engine Driven Generator

Location of generator: ____________________

Location of fuel storage: ____________________ Type of fuel: ____________________

2.3.3 Uninterruptible Power System

Equipment powered by UPS system: ____________________

Location of UPS system: ____________________

Calculated capacity of UPS batteries to drive the system components connected to it:

In standby mode (hours): ____________________ In alarm mode (minutes): ____________________

2.3.4 Batteries

Location: ____________ Type: ________ Nominal voltage: ________ Amp/hour rating: ________

Calculated capacity of batteries to drive the system:

In standby mode (hours): ____________________ In alarm mode (minutes): ____________________

 NFPA 72 (p. 2 of 3)

FIGURE 7.8.2(c) ***Continued***

POWER SYSTEMS
SUPPLEMENTARY RECORD OF COMPLETION *(continued)*

2. SYSTEM POWER *(continued)*

2.4 Supervising Station Transmission Equipment

❑ This system does not use transmission equipment within the building powered by any other source than the alarm system control unit.

2.4.1 Primary Power

Input voltage of shared transmission equipment: ______________________

Shared transmission equipment panel amps: ______________________

Overcurrent protection: Type: ______________ Amps: ______________

Location (of primary supply panelboard): ______________________

Disconnecting means location: ______________________

2.4.2 Engine Driven Generator

Location of generator: ______________________

Location of fuel storage: ______________ Type of fuel: ______________

2.4.3 Uninterruptible Power System

Equipment powered by UPS system: ______________________

Calculated capacity of UPS batteries to drive the system components connected to it:

In standby mode (hours): ______________ In alarm mode (minutes): ______________

2.4.4 Batteries

Location: ______________ Type: ________ Nominal voltage: ________ Amp/hour rating: ________

Calculated capacity of batteries to drive the system:

In standby mode (hours): ______________ In alarm mode (minutes): ______________

See Main System Record of Completion for additional information, certifications, and approvals.

 NFPA 72 (p. 3 of 3)

FIGURE 7.8.2(c) ***Continued***

NOTIFICATION APPLIANCE POWER PANEL SUPPLEMENTARY RECORD OF COMPLETION

This form is a supplement to the System Record of Completion. It includes a list of types and locations of notification appliance power extender panels.

This form is to be completed by the system installation contractor at the time of system acceptance and approval.

It shall be permitted to modify this form as needed to provide a more complete and/or clear record.

Insert N/A in all unused lines.

Form Completion Date: __________ Number of Supplemental Pages Attached: ______

1. PROPERTY INFORMATION

Name of property: ______________________________

Address: ______________________________

2. NOTIFICATION APPLIANCE POWER EXTENDER PANELS

Make and Model	Location	Area Served	Power Source

See Main System Record of Completion for additional information, certifications, and approvals.

 NFPA 72

FIGURE 7.8.2(d) Notification Appliance Power Panel Supplementary Record of Completion. (SIG-FUN)

INTERCONNECTED SYSTEMS SUPPLEMENTARY RECORD OF COMPLETION

This form is a supplement to the System Record of Completion. It includes a list of types and locations of systems that are interconnected to the main system.
This form is to be completed by the system installation contractor at the time of system acceptance and approval.
It shall be permitted to modify this form as needed to provide a more complete and/or clear record.
Insert N/A in all unused lines.

Form Completion Date: ______________ Number of Supplemental Pages Attached: ______

1. PROPERTY INFORMATION

Name of property: ______________________________

Address: ______________________________

2. INTERCONNECTED SYSTEMS

Description	Location	Purpose

See Main System Record of Completion for additional information, certifications, and approvals.

 NFPA 72

FIGURE 7.8.2(e) Interconnected Systems Supplementary Record of Completion. (SIG-FUN)

DEVIATIONS FROM ADOPTED CODES AND STANDARDS SUPPLEMENTARY RECORD OF COMPLETION

This form is a supplement to the System Record of Completion. It enables the designer and/or installer to document and justify deviations from accepted codes or standards.
This form is to be completed by the system installation contractor at the time of system acceptance and approval.
It shall be permitted to modify this form as needed to provide a more complete and/or clear record.
Insert N/A in all unused lines.

Form Completion Date: ________________ Number of Supplemental Pages Attached: __________

1. PROPERTY INFORMATION

Name of property: ________________________________

Address: ________________________________

2. DEVIATIONS FROM ADOPTED CODES OR STANDARDS

Description	Purpose

See Main System Record of Completion for additional information, certifications, and approvals.

 NFPA 72

FIGURE 7.8.2(f) Deviations from Adopted Codes and Standards Supplementary Record of Completion. (SIG-FUN)

SYSTEM RECORD OF INSPECTION AND TESTING

This form is to be completed by the system inspection and testing contractor at the time of a system test.
It shall be permitted to modify this form as needed to provide a more complete and/or clear record.
Insert N/A in all unused lines.
Attach additional sheets, data, or calculations as necessary to provide a complete record.

Inspection/Test Start Date/Time: __________ Inspection/Test Completion Date/Time: __________

Supplemental Form(s) Attached: ________ (yes/no)

1. PROPERTY INFORMATION

Name of property: __________

Address: __________

Description of property: __________

Name of property representative: __________

Address: __________

Phone: __________ Fax: __________ E-mail: __________

2. TESTING AND MONITORING INFORMATION

Testing organization: __________

Address: __________

Phone: __________ Fax: __________ E-mail: __________

Monitoring organization: __________

Address: __________

Phone: __________ Fax: __________ E-mail: __________

Account number: __________ Phone line 1: __________ Phone line 2: __________

Means of transmission: __________

Entity to which alarms are retransmitted: __________ Phone: __________

3. DOCUMENTATION

Onsite location of the required record documents and site-specific software: __________

4. DESCRIPTION OF SYSTEM OR SERVICE

4.1 Control Unit

Manufacturer: __________ Model number: __________

4.2 Software Firmware

Firmware revision number: __________

4.3 System Power

4.3.1 Primary (Main) Power

Nominal voltage: __________ Amps: ______ Location: __________

Overcurrent protection type: __________ Amps: ______ Disconnecting means location: __________

 NFPA 72 (p. 1 of 4)

FIGURE 7.8.2(g) System Record of Inspection and Testing. (SIG-TMS)

SYSTEM RECORD OF INSPECTION AND TESTING *(continued)*

4. DESCRIPTION OF SYSTEM OR SERVICE *(continued)*

4.3.2 Secondary Power

Type: ______________________ Location: ______________________

Battery type (if applicable): ______________________

Calculated capacity of batteries to drive the system:

In standby mode (hours): ______________________ In alarm mode (minutes): ______________________

5. NOTIFICATIONS MADE PRIOR TO TESTING

Monitoring organization	Contact: ______________	Time: ________
Building management	Contact: ______________	Time: ________
Building occupants	Contact: ______________	Time: ________
Authority having jurisdiction	Contact: ______________	Time: ________
Other, if required	Contact: ______________	Time: ________

6. TESTING RESULTS

6.1 Control Unit and Related Equipment

Description	Visual Inspection	Functional Test	Comments
Control unit	❏	❏	
Lamps/LEDs/LCDs	❏	❏	
Fuses	❏	❏	
Trouble signals	❏	❏	
Disconnect switches	❏	❏	
Ground-fault monitoring	❏	❏	
Supervision	❏	❏	
Local annunciator	❏	❏	
Remote annunciators	❏	❏	
Remote power panels	❏	❏	
	❏	❏	

6.2 Secondary Power

Description	Visual Inspection	Functional Test	Comments
Battery condition	❏	❏	
Load voltage	❏	❏	
Discharge test	❏	❏	
Charger test	❏	❏	
Remote panel batteries	❏	❏	

 NFPA 72 (p. 2 of 4)

FIGURE 7.8.2(g) ***Continued***

SYSTEM RECORD OF INSPECTION AND TESTING *(continued)*

6. TESTING RESULTS *(continued)*

6.3 Alarm and Supervisory Alarm Initiating Device

Attach supplementary device test sheets for all initiating devices.

6.4 Notification Appliances

Attach supplementary appliance test sheets for all notification appliances.

6.5 Interface Equipment

Attach supplementary interface component test sheets for all interface components.

Circuit Interface / Signaling Line Circuit Interface / Fire Alarm Control Interface

6.6 Supervising Station Monitoring

Description	Yes	No	Time	Comments
Alarm signal	❑	❑		
Alarm restoration	❑	❑		
Trouble signal	❑	❑		
Trouble restoration	❑	❑		
Supervisory signal	❑	❑		
Supervisory restoration	❑	❑		

6.7 Public Emergency Alarm Reporting System

Description	Yes	No	Time	Comments
Alarm signal	❑	❑		
Alarm restoration	❑	❑		
Trouble signal	❑	❑		
Trouble restoration	❑	❑		
Supervisory signal	❑	❑		
Supervisory restoration	❑	❑		

 NFPA 72 (p. 3 of 4)

FIGURE 7.8.2(g) ***Continued***

SYSTEM RECORD OF INSPECTION AND TESTING *(continued)*

7. NOTIFICATIONS THAT TESTING IS COMPLETE

Monitoring organization Contact: ____________________ Time: __________

Building management Contact: ____________________ Time: __________

Building occupants Contact: ____________________ Time: __________

Authority having jurisdiction Contact: ____________________ Time: __________

Other, if required Contact: ____________________ Time: __________

8. SYSTEM RESTORED TO NORMAL OPERATION

Date: ____________________ Time: __________

9. CERTIFICATION

This system as specified herein has been inspected and tested according to NFPA 72, 2013 edition, Chapter 14.

Signed: ____________________ Printed name: ____________________ Date: __________

Organization: ____________________ Title: ____________________ Phone: __________

Qualifications (refer to 10.5.3): ____________________

10. DEFECTS OR MALFUNCTIONS NOT CORRECTED AT CONCLUSION OF SYSTEM INSPECTION, TESTING, OR MAINTENANCE

10.1 Acceptance by Owner or Owner's Representative:

The undersigned accepted the test report for the system as specified herein:

Signed: ____________________ Printed name: ____________________ Date: __________

Organization: ____________________ Title: ____________________ Phone: __________

 NFPA 72 (p. 4 of 4)

FIGURE 7.8.2(g) ***Continued***

NOTIFICATION APPLIANCE
SUPPLEMENTARY RECORD OF INSPECTION AND TESTING

This form is a supplement to the System Record of Inspection and Testing.
It includes a notification appliance test record.
This form is to be completed by the system inspection and testing contractor at the time of the inspection and/or test.
It shall be permitted to modify this form as needed to provide a more complete and/or clear record.
Insert N/A in all unused lines.

Inspection/Test Start Date/Time: ________________ Inspection/Test Completion Date/Time: ________________

Number of Supplemental Pages Attached: _______

1. PROPERTY INFORMATION

Name of property: ________________

Address: ________________

2. NOTIFICATION APPLIANCE TEST RESULTS

Appliance Type	Location/Identifier	Test Results

 NFPA 72 (p. 1 of 2)

FIGURE 7.8.2(h) Notification Appliance Supplementary Record of Inspection and Testing. (SIG-TMS)

NOTIFICATION APPLIANCE SUPPLEMENTARY RECORD OF INSPECTION AND TESTING *(continued)*

2. NOTIFICATION APPLIANCE TEST RESULTS *(continued)*

Appliance Type	Location/Identifier	Test Results

See main System Record of Inspection and Testing for additional information, certifications, and approvals.

 NFPA 72 (p. 2 of 2)

FIGURE 7.8.2(h) ***Continued***

INITIATING DEVICE
SUPPLEMENTARY RECORD OF INSPECTION AND TESTING

This form is a supplement to the System Record of Inspection and Testing.
It includes an initiating device test record.
This form is to be completed by the system inspection and testing contractor at the time of the inspection and/or test.
It shall be permitted to modify this form as needed to provide a more complete and/or clear record.
Insert N/A in all unused lines.

Inspection/Test Start Date/Time: ________________ Inspection/Test Completion Date/Time: ________________

Number of Supplemental Pages Attached: ________

1. PROPERTY INFORMATION

Name of property: ____

Address: ____

2. INITIATING DEVICE TEST RESULTS

Device Type	Address	Location	Test Results

 NFPA 72 (p. 1 of 2)

FIGURE 7.8.2(i) Initiating Device Supplementary Record of Inspection and Testing. (SIG-TMS)

INITIATING DEVICE
SUPPLEMENTARY RECORD OF INSPECTION AND TESTING *(continued)*

2. INITIATING DEVICE TEST RESULTS ***(continued)***

Device Type	Address	Location	Test Results

See main System Record of Inspection and Testing for additional information, certifications, and approvals.

 NFPA 72 (p. 2 of 2)

FIGURE 7.8.2(i) ***Continued***

MASS NOTIFICATION SYSTEM
SUPPLEMENTARY RECORD OF INSPECTION AND TESTING

This form is a supplement to the System Record of Inspection and Testing.
It includes a mass notification system test record.
This form is to be completed by the system inspection and testing contractor at the time of the inspection and/or test.
It shall be permitted to modify this form as needed to provide a more complete and/or clear record.
Insert N/A in all unused lines.

Inspection/Test Start Date/Time: __________ Inspection/Test Completion Date/Time: __________

Number of Supplemental Pages Attached: ______

1. PROPERTY INFORMATION

Name of property: __________

Address: __________

2. MASS NOTIFICATION SYSTEM

2.1 System Type

❑ In-building MNS—combination

❑ In-building MNS—stand alone ❑ Wide-area MNS ❑ Distributed recipient MNS

❑ Other (specify): __________

2.2 System Features

❑ Combination fire alarm/MNS ❑ MNS ACU only ❑ Wide-area MNS to regional national alerting interface

❑ Local operating console (LOC) ❑ Direct recipient MNS (DRMNS) ❑ Wide-area MNS to DRMNS interface

❑ Wide-area MNS to high-power speaker array (HPSA) interface ❑ In-building MNS to wide-area MNS interface

❑ Other (specify): __________

3. IN-BUILDING MASS NOTIFICATION SYSTEM

3.1 Primary Power

Input voltage of MNS panel: __________ MNS panel amps: __________

3.2 Engine-Driven Generator ❑ This system does not have a generator.

Location of generator: __________

Location of fuel storage: __________ Type of fuel: __________

3.3 Uninterruptible Power System ❑ This system does not have a UPS.

Equipment powered by a UPS system: __________

Location of UPS system: __________

Calculated capacity of UPS batteries to drive the system components connected to it:

In standby mode (hours): __________ In alarm mode (minutes): __________

3.4 Batteries

Location: __________ Type: __________ Nominal voltage: __________ Amp/hour rating: __________

Calculated capacity of batteries to drive the system:

In standby mode (hours): __________ In alarm mode (minutes): __________

❑ Batteries are marked with date of manufacture.

 NFPA 72 (p. 1 of 2)

FIGURE 7.8.2(j) Mass Notification System Supplementary Record of Inspection and Testing. (SIG-TMS)

MASS NOTIFICATION SYSTEM
SUPPLEMENTARY RECORD OF INSPECTION AND TESTING *(continued)*

4. MASS NOTIFICATION EQUIPMENT TEST RESULTS

Description	Visual Inspection	Functional Test	Comments
Functional test			
Reset/power down test			
Fuses			
Primary power supply			
UPS power test			
Trouble signals			
Disconnect switches			
Ground-fault monitoring			
CCU security mechanism			
Prerecorded message content			
Prerecorded message activation			
Software backup performed			
Test backup software			
Fire alarm to MNS interface			
MNS to fire alarm interface			
In-building MNS to wide-area MNS			
MNS to direct recipient MNS			
Sound pressure levels Occupied ❑ Yes ❑ No Ambient dBA:________ Alarm dBA:________ (attach supplementary notification appliance form(s) with locations, values, and weather conditions)			
System intelligibility Test method:____ Score:____ CIS value:________ (attach supplementary notification appliance form(s) with locations, values, and weather conditions)			
Other (specify):			

See main System Record of Inspection and Testing for additional information, certifications, and approvals.

 NFPA 72 (p. 2 of 2)

FIGURE 7.8.2(j) *Continued*

EMERGENCY COMMUNICATIONS SYSTEMS SUPPLEMENTARY RECORD OF INSPECTION AND TESTING

This form is a supplement to the System Record of Inspection and Testing.
It includes systems and components specific to emergency communication systems.
This form is to be completed by the system inspection and testing contractor at the time of the inspection and/or test.
It shall be permitted to modify this form as needed to provide a more complete and/or clear record.
Insert N/A in all unused lines.

Inspection/Test Start Date/Time:__________ Inspection/Test Completion Date/Time:__________

Number of Supplemental Pages Attached: ______

1. PROPERTY INFORMATION

Name of property:__________

Address:__________

2. DESCRIPTION OF SYSTEM OR SERVICE

❑ Fire alarm with in-building fire emergency voice alarm communication system (EVAC)

❑ Mass notification system

❑ Combination system, with the following components:

❑ Fire alarm ❑ EVACS ❑ MNS ❑ Two-way, in-building, emergency communication system

❑ Other (specify):__________

Additional description of system(s):__________

____

2.1 In-Building Fire Emergency Voice Alarm Communication System

Manufacturer:__________ Model number:__________

Number of single voice alarm channels:__________ Number of multiple voice alarm channels:__________

Number of speakers:__________ Number of speaker circuits:__________

Location of amplification and sound processing equipment:__________

____

____

Location of paging microphone stations:

Location 1:__________

Location 2:__________

Location 3:__________

2.2 Mass Notification System

2.2.1 System Type:

❑ In-building MNS—combination

❑ In-building MNS ❑ Wide-area MNS ❑ Distributed recipient MNS

❑ Other (specify):__________

 NFPA 72 (p. 1 of 5)

FIGURE 7.8.2(k) Emergency Communications Systems Supplementary Record of Inspection and Testing. (SIG-TMS)

EMERGENCY COMMUNICATIONS SYSTEMS
SUPPLEMENTARY RECORD OF INSPECTION AND TESTING *(continued)*

2. DESCRIPTION OF SYSTEM OR SERVICE *(continued)*

2.2.2 System Features:

❑ Combination fire alarm/MNS ❑ MNS autonomous control unit ❑ Wide-area MNS to regional national alerting interface

❑ Local operating console (LOC) ❑ Distributed-recipient MNS (DRMNS) ❑ Wide-area MNS to DRMNS interface

❑ Wide-area MNS to high-power speaker array (HPSA) interface ❑ In-building MNS to wide-area MNS interface

❑ Other (specify): ______________________________

2.2.3 MNS Local Operating Consoles

Location 1: ______________________________

Location 2: ______________________________

Location 3: ______________________________

2.2.4 High-Power Speaker Arrays

Number of HPSA speaker initiation zones: ______________________________

Location 1: ______________________________

Location 2: ______________________________

Location 3: ______________________________

2.2.5 Mass Notification Devices

Combination fire alarm/MNS visual devices: ______________ MNS-only visual devices: ______________

Textual signs: ______________ Other (describe): ______________________________

Supervision class: ______________________________

2.2.6 Special Hazard Notification

❑ This system does not have special suppression pre-discharge notification

❑ MNS systems DO NOT override notification appliances required to provide special suppression pre-discharge notification

3. TWO-WAY EMERGENCY COMMUNICATION SYSTEMS

3.1 Telephone System

Number of telephone jacks installed: ______________ Number of warden stations installed: ______________

Number of telephone handsets stored on site: ______________________________

Type of telephone system installed: ❑ Electrically powered ❑ Sound powered

3.2 Area of Refuge (Area of Rescue Assistance) Emergency Communications Systems

Number of stations: ______________ Location of central control point: ______________________________

Days and hours when central control point is attended: ______________________________

Location of alternate control point: ______________________________

Days and hours when alternate control point is attended: ______________________________

 NFPA 72 (p. 2 of 5)

FIGURE 7.8.2(k) *Continued*

EMERGENCY COMMUNICATIONS SYSTEMS SUPPLEMENTARY RECORD OF INSPECTION AND TESTING *(continued)*

3. TWO-WAY EMERGENCY COMMUNICATIONS SYSTEMS *(continued)*

3.3 Elevator Emergency Communications Systems

Number of elevators with stations: ____________ Location of central control point: ____________

Days and hours when central control point is attended: ____________

Location of alternate control point: ____________

Days and hours when alternate control point is attended: ____________

3.4 Other Two-Way Communication System

Describe: ____________

4. TESTING RESULTS

4.1 Control Unit and Related Equipment

Description	Visual Inspection	Functional Test	Comments
Control unit	❑	❑	
Lamps/LEDs/LCDs	❑	❑	
Fuses	❑	❑	
Trouble signals	❑	❑	
Disconnect switches	❑	❑	
Ground fault monitoring	❑	❑	
Supervision	❑	❑	
Local annunciator	❑	❑	
Remote annunciators	❑	❑	
Remote power panels	❑	❑	
Other:	❑	❑	

4.2 Secondary Power

Description	Visual Inspection	Functional Test	Comments
Battery condition	❑	❑	
Load voltage	❑	❑	
Discharge test	❑	❑	
Charger test	❑	❑	
Remote panel batteries	❑	❑	

 NFPA 72 (p. 3 of 5)

FIGURE 7.8.2(k) ***Continued***

EMERGENCY COMMUNICATIONS SYSTEMS SUPPLEMENTARY RECORD OF INSPECTION AND TESTING *(continued)*

4. TESTING RESULTS *(continued)*

4.3 Emergency Communications Equipment

Description	Visual Inspection	Functional Test	Comments
Control unit	❑	❑	
Lamps/LEDs/LCDs	❑	❑	
Fuses	❑	❑	
Secondary power supply	❑	❑	
Trouble signals	❑	❑	
Disconnect switches	❑	❑	
Ground fault monitoring	❑	❑	
Panel supervision	❑	❑	
System performance	❑	❑	
System audibility	❑	❑	
System intelligibility	❑	❑	
Other:	❑	❑	

4.4 Mass Notification Equipment

Description	Visual Inspection	Functional Test	Comments
Functional test	❑	❑	
Reset/Power down test	❑	❑	
Fuses	❑	❑	
Primary power supply	❑	❑	
UPS power test	❑	❑	
Trouble signals	❑	❑	
Disconnect switches	❑	❑	
Ground fault monitoring	❑	❑	
CCU security mechanism	❑	❑	
Prerecorded message content	❑	❑	
Prerecorded message activation	❑	❑	
Software backup performed	❑	❑	
Test backup software	❑	❑	
Fire alarm to MNS Interface	❑	❑	
MNS to fire alarm interface	❑	❑	
In-building MNS to wide-area MNS	❑	❑	
MNS to direct recipient MNS	❑	❑	

 NFPA 72 (p. 4 of 5)

FIGURE 7.8.2(k) ***Continued***

EMERGENCY COMMUNICATIONS SYSTEMS SUPPLEMENTARY RECORD OF INSPECTION AND TESTING *(continued)*

4. TESTING RESULTS *(continued)*

4.4 Mass Notification Equipment *(continued)*

Description	Visual Inspection	Functional Test	Comments
Sound pressure levels (attach report with locations, values, and weather conditions)	❑	❑	
System intelligibility ❑ CSI ❑ STI (attach report with locations, values, and weather conditions)	❑	❑	
Other:	❑	❑	

4.5 Two-Way Communication Equipment

Description	Visual Inspection	Functional Test	Comments
Phone handsets	❑	❑	
Phone jacks	❑	❑	
Off-hook indicator	❑	❑	
Call-in signal	❑	❑	
System performance	❑	❑	
System audibility	❑	❑	
System intelligibility	❑	❑	
Other:	❑	❑	

See main System Record of Inspection and Testing for additional information, certifications, and approvals.

 NFPA 72 (p. 5 of 5)

FIGURE 7.8.2(k) *Continued*

INTERFACE COMPONENT SUPPLEMENTARY RECORD OF INSPECTION AND TESTING

This form is a supplement to the System Record of Inspection and Testing.
It includes an interface component test record for circuit interfaces, signaling line circuit interfaces, and fire alarm control interfaces.
This form is to be completed by the system inspection and testing contractor at the time of the inspection and/or test.
It shall be permitted to modify this form as needed to provide a more complete and/or clear record.
Insert N/A in all unused lines.

Inspection/Test Start Date/Time: ________________ Inspection/Test Completion Date/Time: ________________

Number of Supplemental Pages Attached: ________

1. PROPERTY INFORMATION

Name of property: ________________________________

Address: ________________________________

2. INTERFACE COMPONENT TEST RESULTS

Interface Component Type	Address	Location	Test Results

 NFPA 72 (p. 1 of 2)

FIGURE 7.8.2(l) Interface Component Supplementary Record of Inspection and Testing. (SIG-TMS)

INTERFACE COMPONENT
SUPPLEMENTARY RECORD OF INSPECTION AND TESTING *(continued)*

2. INTERFACE COMPONENT TEST RESULTS *(continued)*

Interface Component Type	Address	Location	Test Results

See main System Record of Inspection and Testing for additional information, certifications, and approvals.

 NFPA 72 (p. 2 of 2)

FIGURE 7.8.2(l) ***Continued***

Chapter 8 Reserved

Chapter 9 Reserved

Chapter 10 Fundamentals

10.1 Application.

10.1.1 The basic functions of a complete fire alarm and/or signaling system shall comply with the requirements of this chapter.

10.1.2 The requirements of this chapter shall apply to systems, equipment, and components addressed in Chapters 12, 14, 17, 18, 21, 23, 24, 26, and 27.

10.1.3 The requirements of Chapter 7 shall apply where referenced in Chapter 10.

10.2 Purpose. The purpose of fire alarm and signaling systems shall be primarily to provide notification of alarm, supervisory, and trouble conditions; to alert the occupants; to summon aid; and to control emergency control functions.

10.3 Equipment.

10.3.1 Equipment constructed and installed in conformity with this Code shall be listed for the purpose for which it is used.

10.3.2 System components shall be installed, tested, inspected, and maintained in accordance with the manufacturer's published instructions and this Code.

10.3.3* All devices and appliances that receive their power from the initiating device circuit or signaling line circuit of a control unit shall be listed for use with the control unit.

10.3.4 All apparatus requiring rewinding or resetting to maintain normal operation shall be restored to normal after each abnormal condition and maintained in normal condition for operation.

10.3.5 Equipment shall be designed so that it is capable of performing its intended functions under the following conditions:

(1)*At 85 percent and at 110 percent of the nameplate primary (main) and secondary (standby) input voltage(s)
(2) At ambient temperatures of 0°C (32°F) and 49°C (120°F)
(3) At a relative humidity of 85 percent and an ambient temperature of 30°C (86°F)

10.4 Design and Installation.

10.4.1* All systems shall be installed in accordance with the plans, specifications, and standards approved by the authority having jurisdiction.

10.4.2 Devices and appliances shall be located and mounted so that accidental operation or failure is not caused by vibration or jarring.

10.4.3 Equipment shall be installed in locations where conditions do not exceed the voltage, temperature, and humidity limits specified in the manufacturer's published instructions.

10.4.4* In areas that are not continuously occupied, an automatic smoke detector shall be provided at the location of each control unit(s), notification appliance circuit power extender(s), and supervising station transmitting equipment to provide notification of fire at that location.

Exception: Where ambient conditions prohibit installation of an automatic smoke detector, an automatic heat detector shall be permitted.

10.4.5 Initiating Devices.

10.4.5.1 Initiating devices of the manual or automatic type shall be selected and installed to minimize unwanted alarms.

10.4.5.2 Initiating devices shall comply with the requirements of Chapter 17 and Chapter 23.

10.4.5.3 Manual alarms shall be initiated by one of the following ways:

(1) A listed manual fire alarm box
(2) A key operated means
(3) A means contained within a locked cabinet or arranged to provide equivalent protection against unauthorized use

10.5 Personnel Qualifications.

10.5.1 System Designer.

10.5.1.1 Fire alarm system and emergency communications system plans and specifications shall be developed in accordance with this Code by persons who are experienced in the design, application, installation, and testing of the systems.

10.5.1.2 State or local licensure regulations shall be followed to determine qualified personnel.

10.5.1.3 Personnel shall provide documentation of their qualification by one or more of the following:

(1) Registration, licensing, or certification by a state or local authority
(2) Certification by an organization acceptable to the authority having jurisdiction
(3) Manufacturer's certification for the specific type and brand of system provided

10.5.1.4 The system designer shall be identified on the system design documents.

10.5.1.5 System design trainees shall be under the supervision of a qualified system designer.

10.5.1.6 The system designer shall provide evidence of their qualifications and/or certifications when required by the authority having jurisdiction.

10.5.2 System Installer.

10.5.2.1 Fire alarm systems and emergency communications systems installation personnel shall be qualified or shall be supervised by persons who are qualified in the installation, inspection, and testing of the systems.

10.5.2.2 State or local licensure regulations shall be followed to determine qualified personnel.

10.5.2.3 Personnel shall provide documentation of their qualification by one or more of the following:

(1) Registration, licensing, or certification by a state or local authority
(2) Certification by an organization acceptable to the authority having jurisdiction
(3) Manufacturer's certification for the specific type and brand of system provided

10.5.2.4 System installation trainees shall be under the supervision of a qualified system installer.

10.5.2.5 The system installer shall provide evidence of their qualifications and/or certifications when requested by the authority having jurisdiction.

10.5.3* Inspection, Testing, and Service Personnel. (SIG-TMS)

10.5.3.1* Inspection Personnel. Inspections shall be performed by personnel who have developed competence through training and experience that are acceptable to the authority having jurisdiction or meet the requirement of 10.5.3.4.

10.5.3.2* Testing Personnel. Testing personnel shall have knowledge and experience of the testing requirements contained in this Code, of the equipment being tested, and of the test methods. That knowledge and experience shall be acceptable to the authority having jurisdiction or meet the requirement of 10.5.3.4.

10.5.3.3 Service Personnel. Service personnel shall have knowledge and experience of the maintenance and servicing requirements contained in this Code, of the equipment being serviced or maintained, and of the servicing or maintenance methods. That knowledge and experience shall be acceptable to the authority having jurisdiction or meet the requirement of 10.5.3.4.

10.5.3.4 Means of Qualification. Qualified personnel shall include, but not be limited to, one or more of the following:

(1)*Personnel who are factory trained and certified for the specific type and brand of system being serviced
(2)*Personnel who are certified by a nationally recognized certification organization acceptable to the authority having jurisdiction
(3)*Personnel, either individually or through their affiliation with an organization that is registered, licensed, or certified by a state or local authority to perform service on systems addressed within the scope of this Code
(4) Personnel who are employed and qualified by an organization listed by a nationally recognized testing laboratory for the servicing of systems within the scope of this Code

10.5.3.5* Programming Personnel.

10.5.3.5.1 Personnel programming a system shall be certified by the system manufacturer.

10.5.3.5.2 System installation personnel shall be permitted to configure systems in the field per manufacturers' published instructions.

10.5.3.5.3 System end users shall be permitted to manage system operation per manufacturers' published instructions or training.

10.5.3.6 Evidence of Qualification. Evidence of qualifications shall be provided to the authority having jurisdiction upon request.

10.5.4 Plans Examiners and Inspectors.

10.5.4.1 Fire alarm system and emergency communications system plans and specifications submitted for review and approval shall be reviewed by personnel who are qualified to review such plans and specifications.

10.5.4.2 Fire alarm system and emergency communications system installations shall be inspected by personnel who are qualified to perform such inspections.

10.5.4.3 State or local licensure regulations shall be followed to determine qualified personnel.

10.5.4.4 Personnel shall provide documentation of their qualifications by one or more of the following:

(1) Registration, licensing, or certification by a state or local authority
(2) Meeting the requirements of NFPA 1031
(3) Assignment by the authority having jurisdiction to perform plan reviews and inspections

10.5.5 Supervising Station Operators. (SIG-SSS)

10.5.5.1 All operators in the supervising station shall demonstrate competence in all tasks required of them in Chapter 26 by one or more of the following:

(1) Certified by the manufacturer of the receiving system or equipment or the alarm-monitoring automation system
(2)*Certified by an organization acceptable to the authority having jurisdiction
(3) Licensed or certified by a state or local authority
(4) Other training or certification approved by the authority having jurisdiction

10.5.5.2 Evidence of qualifications and/or certification shall be provided when requested by the authority having jurisdiction. A license or qualification listing shall be current in accordance with the requirements of the issuing authority or organization.

10.5.5.3 Operator trainees shall be under the direct supervision of a qualified operator until qualified as required by 10.5.5.1.

10.5.6 Public Emergency Alarm Reporting System Personnel Qualification. (SIG-PRS)

10.5.6.1 System Designer.

10.5.6.1.1 Public emergency alarm reporting system plans and specifications shall be developed in accordance with this Code by persons who are qualified in the proper design, application, installation, and testing of public emergency alarm reporting systems.

10.5.6.1.2 The system design documents shall include the name and contact information of the system designer.

10.5.6.2 System Installer. Installation personnel shall be qualified in the installation, inspection, and testing of public emergency alarm reporting systems.

10.5.6.3 Service Personnel. Service personnel shall be qualified in the service, inspection, maintenance, and testing of public emergency alarm reporting systems.

10.5.6.4 Qualification.

10.5.6.4.1 Personnel shall demonstrate qualification by being trained and certified in public emergency alarm reporting system design, installation, or service (as appropriate).

10.5.6.4.2 Personnel who are trained and certified for the specific type of public emergency alarm reporting system and comply with one the following shall be considered qualified:

(1) Personnel who are licensed or certified by a state or local authority, if applicable

(2)*Personnel who are certified by a nationally recognized certification organization acceptable to the authority having jurisdiction
(3) Personnel who are employed and qualified by an organization listed by a nationally recognized testing laboratory for the design, installation, or servicing of systems within the scope of this chapter
(4)*Personnel who are employed and certified by an equipment manufacturer for the specific type of system

10.5.6.4.3 Evidence of qualifications and/or certification shall be provided when requested by the authority having jurisdiction. A license or qualification listing shall be current in accordance with the requirements of the issuing authority or organization.

10.6 Power Supplies.

10.6.1 Scope. The provisions of this section shall apply to power supplies used for protected premises fire alarm systems, supervising station alarm systems, public emergency alarm reporting systems, and emergency communications systems and equipment.

Exception: Low-power radio (wireless) systems that comply with the requirements of 23.16.

10.6.2 Code Conformance. All power supplies shall be installed in accordance with applicable requirements of *NFPA 70*.

10.6.3 Power Supply Sources.

10.6.3.1 Power shall be supplied in accordance with either 10.6.3.2 or 10.6.4.

10.6.3.2 Unless configured in compliance with 10.6.4, at least two independent and reliable power supplies shall be provided, one primary and one secondary.

10.6.3.3 Each power supply shall be of adequate capacity for the application.

10.6.3.4 Monitoring the integrity of power supplies shall be in accordance with 10.6.9.

10.6.4 Uninterruptible Power Supplies (UPS).

10.6.4.1 The UPS device shall be configured in compliance with NFPA 111 for a Type O, Class 24, Level 1 system.

10.6.4.2 The UPS device shall comply with the requirements of 10.6.5.

10.6.4.3 Failure of the UPS shall result in the initiation of a trouble signal in accordance with Section 10.14.

10.6.5 Primary Power Supply.

10.6.5.1 Branch Circuit.

10.6.5.1.1 The branch circuit supplying the fire alarm equipment(s) or emergency communication system(s) shall be supplied by one of the following:

(1) Electric utility
(2) An engine-driven generator or equivalent in accordance with 10.6.11.2, where a person trained in its operation is on duty at all times
(3) An engine-driven generator or equivalent arranged for cogeneration with an electric utility in accordance with 10.6.11.2, where a person trained in its operation is on duty at all times

10.6.5.1.2 The branch circuit supplying the fire alarm equipment or emergency communication system(s) shall supply no other loads.

10.6.5.2 Circuit Identification and Accessibility.

10.6.5.2.1 The location of the branch circuit disconnecting means shall be permanently identified at the control unit.

10.6.5.2.2* The system circuit disconnecting means shall be permanently identified as to its purpose. Methods for marking shall be permitted to include, but not be limited to, one of the following:

(1) "FIRE ALARM" for fire alarm systems
(2) "EMERGENCY COMMUNICATIONS" for emergency communications systems
(3) "FIRE ALARM/ECS" for combination fire alarm and emergency communications systems

10.6.5.2.3 For fire alarm and/or signaling systems, the circuit disconnecting means shall have a red marking.

10.6.5.2.4 The red marking shall not damage the overcurrent protective devices or obscure the manufacturer's markings.

10.6.5.2.5 The circuit disconnecting means shall be accessible only to authorized personnel.

10.6.5.3 Mechanical Protection. The branch circuit(s) and connections shall be protected against physical damage.

10.6.5.4 Circuit Breaker Lock. Where a circuit breaker is the disconnecting means, an approved breaker locking device shall be installed.

10.6.5.5 Overcurrent Protection. An overcurrent protective device shall be provided in accordance with *NFPA 70.*

10.6.6* Continuity of Power Supplies.

10.6.6.1 The secondary power supply shall automatically provide power to the protected premises system within 10 seconds whenever the primary power supply voltage is insufficient for required system operation.

10.6.6.2 The secondary power supply shall automatically provide power to the supervising station facility and equipment within 60 seconds whenever the primary power supply voltage is insufficient for required system operation.

10.6.6.3 Required signals shall not be lost, interrupted, or delayed by more than 10 seconds as a result of the primary power failure.

10.6.6.3.1 Storage batteries dedicated to the system or UPS arranged in accordance with the provisions of NFPA 111 shall be permitted to supplement the secondary power supply to ensure required operation during the transfer period.

10.6.6.3.2 Where a UPS is employed in 10.6.6.3.1, a positive means for disconnecting the input and output of the UPS system while maintaining continuity of power supply to the load shall be provided.

10.6.7 Secondary Power Supply.

10.6.7.1 Secondary Power Operation.

10.6.7.1.1 Operation on secondary power shall not affect the required performance of a system or supervising station facility, including alarm, supervisory, and trouble signals and indications.

Exception: While operating on secondary power, audio amplifier monitoring shall be required only when an alarm is present.

10.6.7.2* Capacity.

10.6.7.2.1 The secondary power supply shall have sufficient capacity to operate the system under quiescent load (system operating in a nonalarm condition) for a minimum of 24 hours and, at the end of that period, shall be capable of operating all alarm notification appliances used for evacuation or to direct aid to the location of an emergency for 5 minutes, unless otherwise permitted or required by 10.6.7.2.1.1 through 10.6.7.2.2:

10.6.7.2.1.1* Battery calculations shall include a minimum 20 percent safety margin above the calculated amp-hour capacity required.

10.6.7.2.1.2 The secondary power supply for in-building fire emergency voice/alarm communications service shall be capable of operating the system under quiescent load for a minimum of 24 hours and then shall be capable of operating the system during a fire or other emergency condition for a period of 15 minutes at maximum connected load.

10.6.7.2.1.3 The secondary power supply capacity for supervising station facilities and equipment shall be capable of supporting operations for a minimum of 24 hours.

10.6.7.2.1.4 The secondary power supply for high-power speaker arrays used for wide-area mass notification systems shall be in accordance with 24.6.5.2.

10.6.7.2.1.5 The secondary power supply for textual visible notification appliances shall be in accordance with 24.6.10.1.

10.6.7.2.1.6 The secondary power supply capacity for emergency command centers of a wide-area mass notification systems shall be capable of supporting operations for a minimum of 24 hours.

10.6.7.2.1.7 The secondary power supply for in-building mass notification systems shall be capable of operating the system under quiescent load for a minimum of 24 hours and then shall be capable of operating the system during emergency conditions for a period of 15 minutes at maximum connected load.

10.6.7.2.2 The secondary power supply capacity required shall include all power supply loads that are not automatically disconnected upon the transfer to secondary power supply.

10.6.7.3* Secondary Power Supply for Protected Premises Fire Alarm Systems and Emergency Communications Systems.

10.6.7.3.1 The secondary power supply shall consist of one of the following:

(1) A storage battery dedicated to the system arranged in accordance with 10.6.10
(2) An automatic-starting, engine-driven generator serving the branch circuit specified in 10.6.5.1 and arranged in accordance with 10.6.11.3.1, and storage batteries dedicated to the system with 4 hours of capacity arranged in accordance with 10.6.10

10.6.7.3.2 Secondary circuits that provide power to the control unit and are not integral to the unit shall be protected against physical damage.

10.6.7.4 Secondary Power Supply for Supervising Station Facilities.

10.6.7.4.1 The secondary power supply shall consist of one of the following:

(1) Storage batteries dedicated to the supervising station equipment arranged in accordance with 10.6.10
(2) A branch circuit of an automatic-starting, engine-driven generator arranged in accordance with 10.6.11.3.2.1 and 10.6.11.3.2.2, and storage batteries dedicated to the supervising station equipment with 4 hours of capacity arranged in accordance with 10.6.10
(3) A branch circuit of multiple engine-driven generators, at least one of which is arranged for automatic starting in accordance with 10.6.11.3.2.1 and 10.6.11.3.2.2

10.6.7.4.2 Where a secondary power supply for supervising station facilities in accordance with 10.6.7.4.1 is used, the following shall apply:

(1) Each generator shall be capable of supplying the energy required.
(2) Generators that are started manually shall be arranged in accordance with 10.6.11.3.2.3 and 10.6.11.3.2.4.
(3) When manual-start generators are employed, a person trained in the procedure of starting the generator shall be on duty at all times.

10.6.8 Power Supply for Remotely Located Control Equipment.

10.6.8.1* Additional power supplies required for system operation shall comply with 10.6.1 through 10.6.6 and with 10.6.9.

10.6.8.2 The location of remotely located power supplies shall be identified at the master control unit.

10.6.8.3 The master control unit display shall be permitted to satisfy the requirement of 10.6.8.2.

10.6.8.4 The location of remotely located power supplies shall be identified on the record drawings.

10.6.9 Monitoring Integrity of Power Supplies.

10.6.9.1 Unless otherwise permitted or required by 10.6.9.1.3 and 10.6.9.1.6, all primary and secondary power supplies shall be monitored for the presence of voltage at the point of connection to the system.

10.6.9.1.1 Failure of either the primary or secondary power supply shall result in a trouble signal in accordance with Section 10.14.

10.6.9.1.2 Power failure indication for a digital alarm communicator transmitter (DACT) powered from a protected premises fire alarm system control unit shall be in accordance with 10.6.9.1.

10.6.9.1.3 Monitoring shall not be required for a power supply for supplementary equipment.

10.6.9.1.4 Monitoring shall not be required for the neutral of a three-, four-, or five-wire ac or dc supply source.

10.6.9.1.5 Monitoring shall not be required for the main power supply in a supervising station if its failure is otherwise indicated and obvious to the operator on duty.

10.6.9.1.6 Monitoring shall not be required for the output of an engine-driven generator that is part of the secondary power supply, provided the generator is tested weekly in accordance with Chapter 14.

10.6.9.2* Power supply sources and electrical supervision for digital alarm communications systems shall be in accordance with Section 10.6, 10.6.9, Section 10.18, and Section 12.6.

10.6.9.3* Supervising station alarm systems shall be arranged to delay transmission of primary power failure signals for 60 minutes to 180 minutes unless a delay is not permitted by the authority having jurisdiction.

10.6.9.4 Power supervisory devices used to monitor the integrity of power supplies shall not impair the receipt of fire alarm or supervisory signals.

10.6.10* Storage Batteries.

10.6.10.1 Marking.

10.6.10.1.1 Batteries shall be marked with the month and year of manufacture using the month/year format.

10.6.10.1.2 Where the battery is not marked with the month/year by the manufacturer, the installer shall obtain the date-code and mark the battery with the month/year of battery manufacture.

10.6.10.2 Arrangement.

10.6.10.2.1 Storage batteries shall comply with the requirements of Article 480 of *NFPA 70.*

10.6.10.2.2 Storage batteries shall be located so that the equipment, including overcurrent devices, is not adversely affected by battery gases.

10.6.10.2.3 Battery cells shall be insulated against ground faults.

10.6.10.2.4 Battery cells shall be insulated to prevent short circuits between multiple cells.

10.6.10.2.5 Battery cells shall be protected from physical damage.

10.6.10.2.6 Battery racks shall be protected against corrosion.

10.6.10.2.7 If not located in or adjacent to the control unit, the batteries and their charger location shall be permanently identified at the control unit.

10.6.10.3 Battery Charging.

10.6.10.3.1 Battery charging equipment shall be provided to keep the battery fully charged under normal conditions.

10.6.10.3.2 Battery charging equipment shall be provided to recharge batteries within 48 hours after fully charged batteries have been subject to a single discharge cycle as specified in 10.6.7.2.

10.6.10.3.3 The battery charging equipment operation shall not damage the battery.

10.6.10.3.4* Batteries shall be either trickle-charged or float-charged.

10.6.10.3.5 Provisions for repair or replacement of failed battery charger equipment shall be maintained at supervising stations and used to restore operation prior to depletion of one-half of the battery capacity.

10.6.10.4 Overcurrent Protection. Overcurrent devices shall be provided to protect the batteries from excessive load current.

10.6.10.5 Metering. The battery charging equipment shall include integral meters or readily accessible terminals so that portable meters can be used to determine battery voltage and charging current.

10.6.10.6 Monitoring Integrity of Battery Charging Equipment.

10.6.10.6.1 Means shall be provided to detect the failure of a battery charger.

10.6.10.6.2 Failure of the battery charger shall result in a trouble signal in accordance with Section 10.14.

10.6.11 Engine-Driven Generators.

10.6.11.1 Application and Installation. The application and installation of engine-driven generators shall be as specified in 10.6.11.2 through 10.6.11.7.

10.6.11.2 Primary Power Supply.

10.6.11.2.1 Engine-driven generators arranged as the primary supply shall be designed in an approved manner.

10.6.11.2.2 Engine-driven generators arranged as the primary supply shall be installed in an approved manner.

10.6.11.3 Secondary Power Supplies.

10.6.11.3.1 Protected Premises.

10.6.11.3.1.1 Engine-driven generators used to provide secondary power for a protected premises fire alarm system or an emergency communications system shall comply with NFPA 110 Chapter 4, requirements for a Type 10, Class 24, Level 1 system.

10.6.11.3.1.2 Installation of engine-driven generators used to provide secondary power for a protected premises fire alarm system or an emergency communications system shall be in accordance with *NFPA 70,* Article 700.

10.6.11.3.1.3 Where survivability of circuits is required by another section of the Code, equal protection shall be provided for power supply circuits.

10.6.11.3.2 Supervising Station.

10.6.11.3.2.1 Automatic-starting, engine-driven generators used to provide secondary power for a supervising station shall comply with NFPA 110 Chapter 4, requirements for a Type 60, Class 24, Level 2 system.

10.6.11.3.2.2 Installation of automatic-starting, engine-driven generators used to provide secondary power for a supervising station shall be in accordance with *NFPA 70* Article 701.

10.6.11.3.2.3 Manual-starting, engine-driven generators used to provide secondary power for a supervising station shall comply with NFPA 110 Chapter 10, requirements for a Type M, Class 24, Level 2 system.

10.6.11.3.2.4 Installation of manual-starting, engine-driven generators used to provide secondary power for a supervising station shall be in accordance with *NFPA 70,* Article 702.

10.6.11.4 Performance, Operation, Testing, and Maintenance. The requirements for performance, operation, testing, and maintenance of engine-driven generators shall conform to the applicable provisions of NFPA 110.

10.6.11.5 Capacity. The unit shall be of a capacity that is sufficient to operate the system under the maximum normal load conditions in addition to all other demands placed upon the unit.

10.6.11.6 Fuel. Unless otherwise required or permitted in 10.6.11.6.1 through 10.6.11.6.3, fuel shall be available in storage sufficient for 6 months of testing plus the capacity specified in 10.6.7.

10.6.11.6.1 For public emergency alarm reporting systems, the requirements of Chapter 27 shall apply.

10.6.11.6.2 If a reliable source of supply is available at any time on a 2-hour notice, it shall be permitted to have fuel in storage sufficient for 12 hours of operation at full load.

10.6.11.6.3 Fuel systems using natural or manufactured gas supplied through reliable utility mains shall not be required to have fuel storage tanks unless located in seismic risk zone 3 or greater as defined in ANSI A-58.1, *Building Code Requirements for Minimum Design Loads in Buildings and Other Structures.*

10.6.11.7 Battery and Charger. A separate storage battery and separate automatic charger shall be provided for starting the engine-driven generator and shall not be used for any other purpose.

10.7 Signal Priority. The priority of signals shall be in accordance with this section.

10.7.1 ECS priority signals when evaluated by stakeholders through a risk analysis in accordance with 24.3.11 shall be permitted to take precedence over all other signals.

10.7.2 Fire alarm signals shall take precedence over all other signals, except as permitted by 10.7.1 or 10.7.3.

10.7.3* Emergency mass notification signals and messages shall be permitted to have priority over fire alarm notification signals in accordance with the requirements of Chapter 24.

10.7.4 Emergency mass notification signals and messages shall have priority over supervisory and trouble signals in accordance with the requirements of Chapter 24.

10.7.5 Carbon monoxide signals shall be permitted to take precedence over supervisory and trouble signals.

10.7.6 Pre-alarm signals shall take precedence over supervisory and trouble signals.

10.7.7 Supervisory signals shall take precedence over trouble signals.

10.7.8 Hold-up alarms or other life-threatening signals shall be permitted to take precedence over supervisory and trouble signals where acceptable to the authority having jurisdiction.

10.7.9* Where separate systems are installed, they shall be permitted to achieve the priority of signals in accordance with Section 10.7.

10.8 Detection and Signaling of Conditions.

10.8.1 Abnormal Condition Detection. Where required by this Code, the system shall be provided with means to detect and signal abnormal conditions.

10.8.2 Alarm Condition Detection. Where required by this Code, the system shall be provided with means to detect and signal alarm conditions.

10.8.2.1 Pre-Alarm Condition Detection. Where required by this Code, the system shall be provided with means to detect and signal pre-alarm conditions.

10.8.2.2 Supervisory Condition Detection. Where required by this Code, the system shall be provided with means to detect and signal supervisory conditions.

10.8.2.3 Trouble Condition Detection. Where required by this Code, the system shall be provided with means to detect and signal trouble conditions.

10.8.2.4 Normal Condition Detection. Where required by this Code, the system shall generate a restoration signal when the device or signaling system returns to normal.

10.9 Responses.

10.9.1 Alarm. The response to an alarm signal shall be in accordance with this Code.

10.9.2 Pre-Alarm. The response to a pre-alarm signal shall be in accordance with this Code.

10.9.3 Supervisory. The response to a supervisory signal shall be in accordance with this Code.

10.9.4 Trouble. The response to trouble signal shall be in accordance with this Code.

10.10 Distinctive Signals.

10.10.1 Priority alarms, fire alarms, supervisory signals, pre-alarm signals, and trouble signals shall be distinctively and descriptively annunciated.

10.10.2 Audible alarm notification appliances for a fire alarm system shall produce signals that are distinctive from other similar appliances used for other purposes in the same area that are not part of the fire alarm or emergency communications system.

10.10.3 Audible alarm notification appliances for a carbon monoxide alarm system shall produce signals that are distinctive from other similar appliances used for other purposes in the same area that are not part of the carbon monoxide, fire alarm, or emergency communications system.

10.10.4* An audible notification appliance on a control unit, on multiple control units that are interconnected to form a system, or at a remote location, shall be permitted to have the same audible characteristics for all alerting functions including, but not limited to, alarm, trouble, and supervisory, provided that the distinction between signals shall be by other means.

10.10.5* Supervisory signals shall be distinctive in sound from other signals, and their sound shall not be used for any other purpose except as permitted in 10.10.4.

10.10.6 Trouble signals required to indicate at the protected premises shall be indicated by distinctive audible signals, which shall be distinctive from alarm signals except as permitted in 10.10.4.

10.10.7 Alarm evacuation signals shall be distinctive in sound from other signals, shall comply with the requirements of 18.4.2, and their sound shall not be used for any other purpose.

10.10.8 Pre-alarm signals shall be distinctive in sound from other signals, and their sound shall not be used for any other purpose except as permitted in 10.10.4.

10.11 Alarm Signals.

10.11.1* Actuation of alarm notification appliances or emergency voice communications, emergency control function interface devices, and annunciation at the protected premises shall occur within 10 seconds after the activation of an initiating device.

10.11.2* Visible notification appliances, textual visible notification appliances, and speaker notification appliances located in the same area shall be activated and deactivated as a group unless otherwise required by an ECS emergency response plan. (SIG-ECS)

10.11.3 Visible alarm strobe notification appliances shall not be activated when speaker notification appliances are used as permitted by 24.3.5 for non-emergency paging. (SIG-ECS)

10.11.4* A coded alarm signal shall consist of not less than three complete rounds of the number transmitted.

10.11.5 Each round of a coded alarm signal shall consist of not less than three impulses.

10.11.6* Resetting of alarm signals shall comply with 23.8.2.2.

10.11.7 The subsequent occurrence of a fault on an initiating device circuit or a signaling line circuit used for other than the interconnection of control units shall not affect previously transmitted unacknowledged alarm signals.

10.11.8 An alarm signal that has been deactivated at the protected premises shall comply with 10.11.8.1 and 10.11.8.2.

10.11.8.1 The audible and visible alarm signal at the control unit only shall automatically reactivate every 24 hours or less until alarm signal conditions are restored to normal.

10.11.8.2 The audible and visible alarm signal shall operate until it is manually silenced or acknowledged.

10.12* Fire Alarm Notification Appliance Deactivation.

10.12.1 A means for turning off activated alarm notification appliance(s) shall be permitted.

10.12.2* When an occupant notification alarm signal deactivation means is actuated, both audible and visible notification appliances shall be simultaneously deactivated.

10.12.3 The fire alarm notification deactivation means shall be key-operated or located within a locked cabinet, or arranged to provide equivalent protection against unauthorized use.

10.12.4 The means shall comply with the requirements of 10.17.1.

10.12.5 Subsequent Actuation of Initiating Devices.

10.12.5.1 Subsequent actuation of nonaddressable initiating devices on other initiating device circuits shall cause the notification appliances to reactivate.

10.12.5.2 Subsequent actuation of addressable alarm initiating devices of a different type in the same room or addressable alarm initiating devices in a different room on signaling line circuits shall cause the notification appliances to reactivate.

10.12.6 A fire alarm notification deactivation means that remains in the deactivated position when there is no alarm condition shall operate an audible trouble notification appliance until the means is restored to normal.

10.13 Supervisory Signals.

10.13.1 Self-Restoring Supervisory Signal Indication. Visible and audible indication of self-restoring supervisory signals and visible indication of their restoration to normal shall be automatically indicated within 90 seconds at the following locations:

(1) Fire alarm control unit for local fire alarm systems
(2) Building fire command center for in-building fire emergency voice/alarm communications systems
(3) Supervising station location for systems installed in compliance with Chapter 26

10.13.2 Latching Supervisory Signal Indication.

10.13.2.1 Visible and audible indication of latching supervisory signals shall be indicated within 90 seconds at the locations specified in 10.13.1.

10.13.2.2 Restoration of latching supervisory signals shall be indicated within 90 seconds at the locations specified in 10.13.1.

10.13.3 Coded Supervisory Signal.

10.13.3.1 A coded supervisory signal shall be permitted to consist of two rounds of the number transmitted to indicate a supervisory off-normal condition.

10.13.3.2 A coded supervisory signal shall be permitted to consist of one round of the number transmitted to indicate the restoration of the supervisory condition to normal.

10.13.4 Combined Coded Alarm and Supervisory Signal Circuits. Where both coded sprinkler supervisory signals and coded fire or waterflow alarm signals are transmitted over the same signaling line circuit, provision shall be made to obtain either alarm signal precedence or sufficient repetition of the alarm signal to prevent the loss of an alarm signal.

10.13.5 Supervisory Notification Appliance Location. The audible supervisory notification appliances shall be located in an area where they are likely to be heard.

10.13.6 Supervisory Signal Reactivation. A supervisory signal that has been deactivated at the protected premises shall comply with 10.13.6.1 and 10.13.6.2.

10.13.6.1 The audible and visible supervisory signal at the control unit only shall automatically reactivate every 24 hours or less until supervisory signal conditions are restored to normal.

10.13.6.2 The audible and visible supervisory signal shall operate until it is manually silenced or acknowledged.

10.13.7 Supervisory Notification Appliance Deactivation.

10.13.7.1 A means for deactivating supervisory notification appliances shall be permitted.

10.13.7.2 The means shall be key-operated or located within a locked cabinet, or arranged to provide equivalent protection against unauthorized use.

10.13.7.3 The means for deactivating supervisory notification appliances shall comply with the requirements of 10.17.2.

10.13.7.4 Subsequent actuation of supervisory initiating devices in other building zones shall cause supervisory notification appliances to actuate as required by the system input/output matrix.

10.13.7.5 A means for deactivating supervisory notification appliances that remains in the deactivated position when there is no supervisory condition shall operate an audible trouble notification appliance until the means is restored to normal.

10.14 Trouble Signals.

10.14.1 Trouble signals and their restoration to normal shall be indicated within 200 seconds at the locations identified in 10.14.7 and 10.14.8.

10.14.2 Indication of primary power failure trouble signals transmitted to a supervising station shall be in accordance with 10.6.9.3.

10.14.3 An audible trouble signal shall be permitted to be intermittent provided it sounds at least once every 10 seconds, with a minimum duration of ½ second.

10.14.4 A single audible trouble signal shall be permitted to annunciate multiple fault conditions.

10.14.5 The audible trouble notification appliances shall be located in an area where they are likely to be heard.

10.14.6 Actuated notification appliances at the protected premises shall continue to operate unless they are manually silenced as permitted by 10.14.10.1.

10.14.7 Visible and audible trouble signals and visible indication of their restoration to normal shall be indicated at the following locations:

(1) Fire alarm control unit for protected premises alarm systems
(2) Building fire command center for in-building fire emergency voice/alarm communications systems
(3) Central station or remote station location for systems installed in compliance with Chapter 26

10.14.8 Trouble signals and their restoration to normal shall be visibly and audibly indicated at the proprietary supervising station for systems installed in compliance with Chapter 26.

10.14.9* A trouble signal that has been deactivated at the protected premises shall comply with 10.14.9.1 and 10.14.9.2.

10.14.9.1 The audible and visible trouble signal shall automatically reactuate at the control unit every 24 hours or less until trouble signal conditions are restored to normal.

10.14.9.2 The audible and visible trouble signal associated with signaling the depletion or failure of the primary battery of a wireless system as required by 23.16.2(3) and (4) shall automatically resound every 4 hours or less until the depletion signal is restored to normal.

10.14.10 Trouble Notification Appliance Deactivation.

10.14.10.1 A means for deactivating trouble notification appliances shall be permitted.

10.14.10.2 The means shall be key-operated or located within a locked cabinet, or arranged to provide equivalent protection against unauthorized use.

10.14.10.3 The means for deactivating trouble notification appliances shall comply with the requirements of 10.17.2.

10.14.10.4 If an audible trouble notification appliance is also used to indicate a supervisory condition, as permitted by 10.10.4, a trouble notification appliance deactivation means shall not prevent subsequent actuation of supervisory notification appliances.

10.14.10.5 Subsequent trouble signals shall cause trouble notification appliances to activate as required by the system input/output matrix.

10.14.10.6 A means for deactivating trouble notification appliances that remains in the deactivated position when there is no trouble condition shall operate an audible trouble notification appliance until the means is restored to normal.

10.14.10.7* Unless otherwise permitted by the authority having jurisdiction, trouble notification appliances at the protected premises of a supervising station fire alarm system arranged in accordance with Chapter 26, that have been silenced at the protected premises shall automatically reactuate every 24 hours or less until fault conditions are restored to normal.

10.15 Emergency Control Function Status Indicators.

10.15.1 All controls provided specifically for the purpose of manually overriding any automatic emergency control function shall provide visible indication of the status of the associated control circuits.

10.15.2* Where status indicators are provided for emergency equipment or control functions, they shall be arranged to reflect the actual status of the associated equipment or function.

10.16 Notification Appliance Circuits and Control Circuits.

10.16.1 An open, ground-fault, or short-circuit fault on the installation conductors of one alarm notification appliance circuit shall not affect the operation of any other alarm notification appliance circuit for more than 200 seconds regardless of whether the short-circuit fault is present during the normal or activated circuit state.

10.16.2* Notification appliance circuits that do not have notification appliances connected directly to the circuit shall be considered control circuits.

10.16.3 Control circuits shall not be required to comply with 10.16.1, provided that the circuit is monitored for integrity in accordance with Section 12.6 and a fault in installation conductors shall result in a trouble signal in accordance with Section 10.14.

10.17 Annunciation and Annunciation Zoning.

10.17.1 Alarm Annunciation.

10.17.1.1 Where required by other governing laws, codes, or standards, the location of an operated initiating device shall be annunciated by visible means.

10.17.1.1.1 Visible annunciation of the location of an operated initiating device shall be by an indicator lamp, alphanumeric display, printout, or other approved means.

10.17.1.1.2 The visible annunciation of the location of operated initiating devices shall not be canceled by the means used to deactivate alarm notification appliances.

10.17.2 Supervisory and Trouble Annunciation.

10.17.2.1 Where required by other governing laws, codes, or standards, supervisory and/or trouble conditions shall be annunciated by visible means.

10.17.2.1.1 Visible annunciation shall be by an indicator lamp, an alphanumeric display, a printout, or other means.

10.17.2.1.2 The visible annunciation of supervisory and/or trouble conditions shall not be canceled by the means used to deactivate supervisory or trouble notification appliances.

10.17.3* Annunciator Access and Location.

10.17.3.1 All required annunciation means shall be readily accessible to responding personnel.

10.17.3.2 All required annunciation means shall be located as required by the authority having jurisdiction to facilitate an efficient response to the situation.

10.17.4 Alarm Annunciation Display. Visible annunciators shall be capable of displaying all zones in alarm.

10.17.4.1 If all zones in alarm are not displayed simultaneously, the zone of origin shall be displayed.

10.17.4.2 If all zones in alarm are not displayed simultaneously, there shall be an indication that other zones are in alarm.

10.17.5* Annunciation Zoning.

10.17.5.1 For the purpose of alarm annunciation, each floor of the building shall be considered as a separate zone.

10.17.5.2 For the purposes of alarm annunciation, if a floor of the building is subdivided into multiple zones by fire or smoke barriers and the fire plan for the protected premises allows relocation of occupants from the zone of origin to another zone on the same floor, each zone on the floor shall be annunciated separately.

10.17.5.3 Where the system serves more than one building, each building shall be annunciated separately.

10.18 Monitoring Integrity of In-Building Fire Emergency Voice/Alarm Communications Systems.

10.18.1* Speaker Amplifier and Tone-Generating Equipment. If speakers are used to produce audible fire alarm signals, the required trouble signal for 10.18.1.1 through 10.18.1.3 shall be in accordance with Section 10.14.

10.18.1.1 When primary power is available, failure of any audio amplifier shall result in a trouble signal.

10.18.1.2 When an alarm is present and primary power is not available (i.e., system is operating from the secondary power source), failure of any audio amplifier shall result in a trouble signal.

10.18.1.3 Failure of any tone-generating equipment shall result in a trouble signal, unless the tone-generating and amplifying equipment are enclosed as integral parts and serve only a single, listed loudspeaker.

10.18.2 Two-Way Telephone Communications Circuits.

10.18.2.1 Two-way telephone communications circuit installation conductors shall be monitored for open circuit fault conditions that would cause the telephone communications circuit to become fully or partially inoperative.

10.18.2.2 Two-way telephone communications circuit installation conductors shall be monitored for short circuit fault conditions that would cause the telephone communications circuit to become fully or partially inoperative.

10.18.2.3 Two-way telephone communications circuit fault conditions shall result in a trouble signal in accordance with Section 10.14.

10.19 Documentation and Notification.

10.19.1 Documentation shall be in accordance with Chapter 7.

10.19.2 The authority having jurisdiction shall be notified prior to installation or alteration of equipment or wiring.

10.20* Impairments.

10.20.1 The system owner or the owner's designated representative shall be notified when a system or part thereof is impaired. Impairments to systems shall include out-of-service events.

10.20.2 A record of the impairments shall be maintained by the system owner or the owner's designated representative for a period of 1 year from the date the impairment is corrected.

10.20.3 The supervising station shall report to the authority having jurisdiction any system for which required monitoring has been terminated.

10.20.4* The service provider shall report to the authority having jurisdiction any system that is out of service for more than 8 hours.

10.20.5* Where required by the authority having jurisdiction, mitigating measures shall be implemented for the period that the system is impaired.

10.20.6 The system owner or the owner's designated representative and the authority having jurisdiction shall be notified when an impairment period ends.

10.21* Unwanted Alarms. For the purpose of reporting, alarm signals that are not the result of hazardous conditions shall be classified as Unwanted and subclassified as one of the following:

(1) Malicious alarm
(2) Nuisance alarm
(3) Unintentional alarm
(4) Unknown alarm

Chapter 11 Reserved

Chapter 12 Circuits and Pathways

12.1 Application.

12.1.1 Pathways (interconnections) shall be designated based on the performance characteristics defined in this chapter.

12.1.2 The requirements of Chapter 14 shall apply.

12.2 General.

12.2.1* Performance and survivability characteristics of signaling pathways (interconnections) shall comply with the defined designations of this chapter.

12.2.2 A pathway (interconnection) class designation shall be dependent on the pathway (interconnection) capability to continue to operate during abnormal conditions.

12.2.3 The installation of all pathway wiring, cable, and equipment shall be in accordance with *NFPA 70* and the applicable requirements of 12.2.3.1 through 12.2.3.3.

12.2.3.1 Optical fiber cables installed as part of the fire alarm system shall meet the requirements of *NFPA 70* Article 770, and be protected against physical damage in accordance with *NFPA 70* Article 760.

12.2.3.2* Fire alarm system wiring and equipment, including all circuits controlled and powered by the fire alarm system, shall be installed in accordance with the requirements of this Code and of *NFPA 70* Article 760.

12.2.3.3* Wiring methods permitted by other sections of this Code to resist attack by fire shall be installed in accordance with manufacturer's published instructions and the requirements of *NFPA 70* Article 760.

12.2.3.4* Where operational capability is required to be maintained or continued during the application of a fault, the operational capability required in 10.11.1 shall be restored within 200 seconds from the time the fault is introduced.

12.2.4 Ground Connections.

12.2.4.1 All fire alarm systems shall test free of grounds.

Exception: Parts of circuits or equipment that are intentionally and permanently grounded to provide ground-fault detection, noise suppression, emergency ground signaling, and circuit protection grounding shall be permitted.

12.2.4.2* On conductive pathways, operational capability shall be maintained during the application of a single ground connection.

12.3* Pathway Class Designations. Pathways shall be designated as Class A, Class B, Class C, Class D, Class E, Class N, or Class X, depending on their performance.

12.3.1* Class A. A pathway shall be designated as Class A when it performs as follows:

(1) It includes a redundant path.
(2) Operational capability continues past a single open, and the single open fault shall result in the annunciation of a trouble signal.
(3) Conditions that affect the intended operation of the path are annunciated as a trouble signal.
(4) Operational capability is maintained during the application of a single ground fault.
(5) A single ground condition shall result in the annunciation of a trouble signal.

Exception: Requirements in 12.3.1(4) and (5) shall not apply to nonconductive pathways (e.g., wireless or fiber).

12.3.2 Class B. A pathway shall be designated as Class B when it performs as follows:

(1) It does not include a redundant path.
(2) Operational capability stops at a single open.
(3) Conditions that affect the intended operation of the path are annunciated as a trouble signal.
(4) Operational capability is maintained during the application of a single ground fault.
(5) A single ground condition shall result in the annunciation of a trouble signal.

Exception: Requirements in 12.3.2(4) and (5) shall not apply to nonconductive pathways (e.g., wireless or fiber).

12.3.3* Class C. A pathway shall be designated as Class C when it performs as follows:

(1) It includes one or more pathways where operational capability is verified via end-to-end communication, but the integrity of individual paths is not monitored.
(2) A loss of end-to-end communication is annunciated as a trouble signal.

12.3.4* Class D. A pathway shall be designated as Class D when it has fail-safe operation, where no fault is annunciated, but the intended operation is performed in the event of a pathway failure.

12.3.5* Class E. A pathway shall be designated as Class E when it is not monitored for integrity.

12.3.6 Class N. A pathway shall be designated as Class N when it performs as follows:

(1)*It includes two or more pathways where operational capability of the primary pathway and a redundant pathway to each device shall be verified through end-to-end communication.

Exception: When only one device is served, only one pathway shall be required.

(2) A loss of intended communications between endpoints shall be annunciated as a trouble signal.
(3) A single open, ground, short, or combination of faults on one pathway shall not affect any other pathway.
(4)*Conditions that affect the operation of the primary pathway(s) and redundant pathway(s) shall be annunciated as a trouble signal when the system's minimal operational requirements cannot be met.
(5)*Primary and redundant pathways shall not be permitted to share traffic over the same physical segment.

12.3.7 Class X. A pathway shall be designated as Class X when it performs as follows:

(1) It includes a redundant path.
(2) Operational capability continues past a single open, and the single open fault shall result in the annunciation of a trouble signal.
(3) Operational capability continues past a single short-circuit, and the single short-circuit fault shall result in the annunciation of a trouble signal.
(4) Operational capability continues past a combination open fault and ground fault.
(5) Conditions that affect the intended operation of the path are annunciated as a trouble signal.
(6) Operational capability is maintained during the application of a single ground fault.
(7) A single ground condition shall result in the annunciation of a trouble signal.

Exception: Requirements in 12.3.7(3), (4), (6), and (7) shall not apply to nonconductive pathways (e.g., wireless or fiber).

12.3.8* Class A and Class X Pathway Separation. Class A and Class X circuits using physical conductors (e.g., metallic, optical fiber) shall be installed so that the primary and redundant, or outgoing and return conductors, exiting from and returning to the control unit, respectively, are routed separately.

12.3.8.1 The outgoing and return (redundant) circuit conductors shall be permitted in the same cable assembly (i.e., multiconductor cable), enclosure, or raceway only under the following conditions:

(1) For a distance not to exceed 10 ft (3.0 m) where the outgoing and return conductors enter or exit the initiating device, notification appliance, or control unit enclosures
(2) Single drops installed in the raceway to individual devices or appliances
(3)*In a single room not exceeding 1000 ft^2 (93 m^2) in area, a drop installed in the raceway to multiple devices or appliances that does not include any emergency control function devices

12.4* Pathway Survivability. All pathways shall comply with *NFPA 70.*

12.4.1 Pathway Survivability Level 0. Level 0 pathways shall not be required to have any provisions for pathway survivability.

12.4.2 Pathway Survivability Level 1. Pathway survivability Level 1 shall consist of pathways in buildings that are fully protected by an automatic sprinkler system in accordance with NFPA 13 with any interconnecting conductors, cables, or other physical pathways installed in metal raceways.

12.4.3* Pathway Survivability Level 2. Pathway survivability Level 2 shall consist of one or more of the following:

(1) 2-hour fire-rated circuit integrity (CI) or fire-resistive cable
(2) 2-hour fire-rated cable system [electrical circuit protective system(s)]
(3) 2-hour fire-rated enclosure or protected area
(4)*Performance alternatives approved by the authority having jurisdiction

12.4.4 Pathway Survivability Level 3. Pathway survivability Level 3 shall consist of pathways in buildings that are fully protected by an automatic sprinkler system in accordance with NFPA 13 and one or more of the following:

(1) 2-hour fire-rated circuit integrity (CI) or fire-resistive cable
(2) 2-hour fire-rated cable system [electrical circuit protective system(s)]
(3) 2-hour fire-rated enclosure or protected area
(4)*Performance alternatives approved by the authority having jurisdiction

12.5* Shared Pathway Designations. Shared pathways shall be designated as Level 0, Level 1, Level 2, or Level 3, depending on their performance.

12.5.1* Shared Pathway Level 0. Level 0 pathways shall not be required to segregate or prioritize life safety data from non–life safety data.

12.5.2* Shared Pathway Level 1. Level 1 pathways shall not be required to segregate life safety data from non–life safety data, but shall prioritize all life safety data over non–life safety data.

12.5.3* Shared Pathway Level 2. Level 2 pathways shall segregate all life safety data from non–life safety data.

12.5.4* Shared Pathway Level 3. Level 3 pathways shall use equipment that is dedicated to the life safety system.

12.6* Monitoring Integrity and Circuit Performance of Installation Conductors and Other Signaling Channels.

12.6.1 Unless otherwise permitted or required by 12.3.1 through 12.3.7 and 12.6.3 through 12.6.13, all means of interconnecting equipment, devices, and appliances and wiring connections shall be monitored for the integrity of the interconnecting conductors or equivalent path so that the occurrence of a single open or a single ground-fault condition in the installation conductors or other signaling channels is automatically indicated within 200 seconds.

12.6.2 Unless otherwise permitted or required by 12.3.1 through 12.3.7 and 12.6.3 through 12.6.13, all means of interconnecting equipment, devices, and appliances and wiring connections shall be monitored for the integrity of the interconnecting conductors or equivalent path so that the restoration to normal of a single open or a single ground-fault condition in the installation conductors or other signaling channels is automatically indicated within 200 seconds.

12.6.3 Shorts between conductors shall not be required to be monitored for integrity, unless required by 12.6.15, 12.6.16, and 10.18.2.

12.6.4 Monitoring for integrity shall not be required for a noninterfering shunt circuit, provided that a fault circuit condition on the shunt circuit wiring results only in the loss of the noninterfering feature of operation.

12.6.5 Monitoring for integrity shall not be required for connections to and between supplementary system components, provided that a single open, ground-fault, or short-circuit conditions of the supplementary equipment or interconnecting means, or both, do not affect the required operation of the fire alarm and/or signaling system.

12.6.6 Monitoring for integrity shall not be required for the circuit of an alarm notification appliance installed in the same room with the central control equipment, provided that the notification appliance circuit conductors are installed in conduit or are equivalently protected against mechanical injury.

12.6.7 Monitoring for integrity shall not be required for a trouble notification appliance circuit.

12.6.8* Monitoring for integrity shall not be required for the interconnection between listed equipment within a common enclosure.

12.6.9 Monitoring for integrity shall not be required for the interconnection between enclosures containing control equipment located within 20 ft (6 m) of each other where the conductors are installed in conduit or equivalently protected against mechanical injury.

12.6.10 Monitoring for integrity shall not be required for the conductors for ground-fault detection where a single ground-fault does not prevent the required normal operation of the system.

12.6.11 Monitoring for integrity shall not be required for pneumatic rate-of-rise systems of the continuous line type in which the wiring terminals of such devices are connected in multiple across electrically supervised circuits.

12.6.12 Monitoring for integrity shall not be required for the interconnecting wiring of a stationary computer and the computer's keyboard, video monitor, mouse-type device, or touch screen, as long as the interconnecting wiring does not exceed 8 ft (2.4 m) in length; is a listed computer/data processing cable as permitted by *NFPA 70*; and failure of cable does not cause the failure of the required system functions not initiated from the keyboard, mouse, or touch screen.

12.6.13 Monitoring for integrity of the installation conductors for a ground-fault condition shall not be required for the communications and transmission channels extending from a supervising station to a subsidiary station(s) or protected premises, or both, that comply with the requirements of Chapter 26 and are electrically isolated from the fire alarm system (or circuits) by a transmitter(s).

12.6.14 Interconnection means shall be arranged so that a single break or single ground-fault does not cause an alarm signal.

12.6.15 A wire-to-wire short-circuit fault on any alarm notification appliance circuit shall result in a trouble signal in accordance with Section 10.14, except as permitted by 12.6.5, 12.6.6, or 12.6.11.

12.6.16 Where two or more systems are interconnected, the systems shall be connected using Class A, B, N, or X circuits as described in Section 12.3.

12.7* Nomenclature. The following nomenclature shall be used to identify the required properties of the system(s) interconnections and survivability:

(1) System(s) interconnections
(2) Survivability levels (not required if Level 0)
(3) Shared pathway levels (not required if Level 0)

Chapter 13 Reserved

Chapter 14 Inspection, Testing, and Maintenance

14.1 Application.

14.1.1 The inspection, testing, and maintenance of systems, their initiating devices, and notification appliances shall comply with the requirements of this chapter.

14.1.2 The inspection, testing, and maintenance of single- and multiple-station smoke and heat alarms and household fire alarm systems shall comply with the requirements of this chapter.

14.1.3 Procedures that are required by other parties and that exceed the requirements of this chapter shall be permitted.

14.1.4 The requirements of this chapter shall apply to both new and existing systems.

14.1.5 The requirements of Chapter 7 shall apply where referenced in Chapter 14.

14.2 General.

14.2.1 Purpose.

14.2.1.1* The purpose for initial and reacceptance inspections is to ensure compliance with approved design documents and to ensure installation in accordance with this Code and other required installation standards.

14.2.1.2* The purpose for initial and reacceptance tests of fire alarm and signaling systems is to ensure system operation in accordance with the design documents.

14.2.1.3* The purpose for periodic inspections is to assure that obvious damages or changes that might affect the system operability are visually identified.

14.2.1.4* The purpose for periodic testing is to statistically assure operational reliability.

14.2.2 Performance.

14.2.2.1 Performance Verification. To ensure operational integrity, the system shall have an inspection, testing, and maintenance program.

14.2.2.1.1 Inspection, testing, and maintenance programs shall satisfy the requirements of this Code and conform to the equipment manufacturer's published instructions.

14.2.2.1.2 Inspection, testing, and maintenance programs shall verify correct operation of the system.

14.2.2.2 Impairments/Deficiencies.

14.2.2.2.1 The requirements of Section 10.20 shall be applicable when a system is impaired.

14.2.2.2.2 System deficiencies shall be corrected.

14.2.2.2.3 If a deficiency is not corrected at the conclusion of system inspection, testing, or maintenance, the system owner or the owner's designated representative shall be informed of the impairment in writing within 24 hours.

14.2.2.2.4 In the event that any equipment is observed to be part of a recall program, the system owner or the system owner's designated representative shall be notified in writing.

14.2.3 Responsibilities.

14.2.3.1* The property or building or system owner or the owner's designated representative shall be responsible for inspection, testing, and maintenance of the system and for alterations or additions to this system.

14.2.3.2 Where the property owner is not the occupant, the property owner shall be permitted to delegate the authority and responsibility for inspecting, testing, and maintaining the fire protection systems to the occupant, management firm, or managing individual through specific provisions in the lease, written use agreement, or management contract.

14.2.3.3 Inspection, testing, or maintenance shall be permitted to be done by the building or system owner or a person or organization other than the building or system owner if conducted under a written contract.

14.2.3.4 Where the building or system owner has delegated any responsibilities for inspection, testing, or maintenance, a copy of the written delegation required by 14.2.3.3 shall be provided to the authority having jurisdiction upon request.

14.2.3.5 Testing and maintenance of central station service systems shall be performed under the contractual arrangements specified in 26.3.3.

14.2.3.6* Service Personnel Qualifications and Experience. Service personnel shall be qualified and experienced in accordance with the requirements of 10.5.3.

14.2.4* Notification.

14.2.4.1 Before proceeding with any testing, all persons and facilities receiving alarm, supervisory, or trouble signals and all building occupants shall be notified of the testing to prevent unnecessary response.

14.2.4.2 At the conclusion of testing, those previously notified (and others, as necessary) shall be notified that testing has been concluded.

14.2.4.3 The owner or the owner's designated representative and service personnel shall coordinate system testing to prevent interruption of critical building systems or equipment.

14.2.5 System Documentation. Prior to system maintenance or testing, the record of completion and any information required by Chapter 7 regarding the system and system alterations, including specifications, wiring diagrams, and floor plans, shall be provided by the owner or a designated representative to the service personnel upon request.

14.2.5.1 The provided documentation shall include the current revisions of all fire alarm software and the revisions of software of any systems with which the fire alarm software interfaces.

14.2.5.2 The revisions of fire alarm software, and the revisions of the software in the systems with which the fire alarm software interfaces, shall be verified for compatibility in accordance with the requirements of 23.2.2.1.1.

14.2.6 Releasing Systems. Requirements pertinent to testing the fire alarm systems initiating fire suppression system releasing functions shall be covered by 14.2.6.1 through 14.2.6.6.

14.2.6.1 Testing personnel shall be qualified and experienced in the specific arrangement and operation of a suppression system(s) and a releasing function(s) and shall be cognizant of the hazards associated with inadvertent system discharge.

14.2.6.2 Occupant notification shall be required whenever a fire alarm system configured for releasing service is being serviced or tested.

14.2.6.3 Discharge testing of suppression systems shall not be required by this Code.

14.2.6.4 Suppression systems shall be secured from inadvertent actuation, including disconnection of releasing solenoids or electric actuators, closing of valves, other actions, or combinations thereof, for the specific system, for the duration of the fire alarm system testing.

14.2.6.5 Testing shall include verification that the releasing circuits and components energized or actuated by the fire alarm system are electrically monitored for integrity and operate as intended on alarm.

14.2.6.6 Suppression systems and releasing components shall be returned to their functional operating condition upon completion of system testing.

14.2.7 Interface Equipment and Emergency Control Functions.

14.2.7.1* Testing personnel shall be qualified and experienced in the arrangement and operation of interface equipment and emergency control functions.

14.2.7.2 Testing shall be accomplished in accordance with Table 14.4.3.2.

14.2.8 Automated Testing.

14.2.8.1 Automated testing arrangements that provide equivalent means of testing devices to those specified in Table 14.4.3.2 at a frequency at least equivalent to those specified in Table 14.4.3.2 shall be permitted to be used to comply with the requirements of this chapter.

14.2.8.2 Failure of a device on an automated test shall result in an audible and visual trouble signal.

14.2.9* Performance-Based Inspection and Testing. As an alternate means of compliance, subject to the authority having jurisdiction, components and systems shall be permitted to be inspected and tested under a performance-based program.

14.2.10* Test Plan.

14.2.10.1 A test plan shall be written to clearly establish the scope of the testing for the fire alarm or signaling system.

14.2.10.2 The test plan and results shall be documented with the testing records.

14.3 Inspection.

14.3.1* Unless otherwise permitted by 14.3.2, visual inspections shall be performed in accordance with the schedules in Table 14.3.1 or more often if required by the authority having jurisdiction.

Table 14.3.1 Visual Inspection

Component	Initial Acceptance	Periodic Frequency	Method	Reference
1. All equipment	X	Annual	Ensure there are no changes that affect equipment performance. Inspect for building modifications, occupancy changes, changes in environmental conditions, device location, physical obstructions, device orientation, physical damage, and degree of cleanliness.	14.3.4
2. Control equipment:				
(a) Fire alarm systems monitored for alarm, supervisory, and trouble signals			Verify a system normal condition.	
(1) Fuses	X	Annual		
(2) Interfaced equipment	X	Annual		
(3) Lamps and LEDs	X	Annual		
(4) Primary (main) power supply	X	Annual		
(5) Trouble signals	X	Semiannual		
(b) Fire alarm systems unmonitored for alarm, supervisory, and trouble signals			Verify a system normal condition.	
(1) Fuses	X	Weekly		
(2) Interfaced equipment	X	Weekly		
(3) Lamps and LEDs	X	Weekly		
(4) Primary (main) power supply	X	Weekly		
(5) Trouble signals	X	Weekly		

Table 14.3.1 ***Continued***

Component	Initial Acceptance	Periodic Frequency	Method	Reference
3. Reserved				
4. Supervising station alarm systems — transmitters			Verify location, physical condition, and a system normal condition.	
(a) Digital alarm communicator transmitter (DACT)	X	Annual		
(b) Digital alarm radio transmitter (DART)	X	Annual		
(c) McCulloh	X	Annual		
(d) Radio alarm transmitter (RAT)	X	Annual		
(e) All other types of communicators	X	Annual		
5. In-building fire emergency voice/alarm communications equipment	X	Semiannual	Verify location and condition.	
6. Reserved				
7. Reserved				
8. Reserved				
9. Batteries			Inspect for corrosion or leakage. Verify tightness of connections. Verify marking of the month/year of manufacture (all types).	10.6.10
(a) Lead-acid	X	Monthly	Visually inspect electrolyte level.	
(b) Nickel-cadmium	X	Semiannual		
(c) Primary (dry cell)	X	Monthly		
(d) Sealed lead-acid	X	Semiannual		
10. Reserved				
11. Remote annunciators	X	Semiannual	Verify location and condition.	
12. Notification appliance circuit power extenders	X	Annual	Verify proper fuse ratings, if any. Verify that lamps and LEDs indicate normal operating status of the equipment.	10.6
13. Remote power supplies	X	Annual	Verify proper fuse ratings, if any. Verify that lamps and LEDs indicate normal operating status of the equipment.	10.6
14. Transient suppressors	X	Semiannual	Verify location and condition.	
15. Reserved				
16. Fiber-optic cable connections	X	Annual	Verify location and condition.	
17. Initiating devices			Verify location and condition (all devices).	
(a) Air sampling				
(1) General	X	Semiannual	Verify that in-line filters, if any, are clean.	17.7.3.6
(2) Sampling system piping and sampling ports	X	N/A	Verify that sampling system piping and fittings are installed properly, appear airtight, and are permanently fixed. Confirm that sampling pipe is conspicuously identified. Verify that sample ports or points are not obstructed.	17.7.3.6

(continues)

Table 14.3.1 ***Continued***

Component	Initial Acceptance	Periodic Frequency	Method	Reference
(b) Duct detectors				
(1) General	X	Semiannual	Verify that detector is rigidly mounted. Confirm that no penetrations in a return air duct exist in the vicinity of the detector. Confirm the detector is installed so as to sample the airstream at the proper location in the duct.	17.7.5.5
(2) Sampling tube	X	Annual	Verify proper orientation. Confirm the sampling tube protrudes into the duct in accordance with system design.	17.7.5.5
(c) Electromechanical releasing devices	X	Semiannual		
(d) Fire extinguishing system(s) or suppression system(s) switches	X	Semiannual		
(e) Manual fire alarm boxes	X	Semiannual		
(f) Heat detectors	X	Semiannual		
(g) Radiant energy fire detectors	X	Quarterly	Verify no point requiring detection is obstructed or outside the detector's field of view.	17.8
(h) Video image smoke and fire detectors	X	Quarterly	Verify no point requiring detection is obstructed or outside the detector's field of view.	17.7.7; 17.11.5
(i) Smoke detectors (excluding one- and two-family dwellings)	X	Semiannual		
(j) Projected beam smoke detectors	X	Semiannual	Verify beam path is unobstructed.	
(k) Supervisory signal devices	X	Quarterly		
(l) Waterflow devices	X	Quarterly		
18. Reserved				
19. Combination systems			Verify location and condition (all types).	
(a) Fire extinguisher electronic monitoring device/systems	X	Semiannual		
(b) Carbon monoxide detectors/systems	X	Semiannual		
20. Fire alarm control interface and emergency control function interface	X	Semiannual	Verify location and condition.	
21. Guard's tour equipment	X	Semiannual	Verify location and condition.	
22. Notification appliances			Verify location and condition (all appliances).	
(a) Audible appliances	X	Semiannual		
(b) Audible textual notification appliances	X	Semiannual		
(c) Visible appliances				
(1) General	X	Semiannual		18.5.5
(2) Candela rating	X	N/A	Verify that the candela rating marking agrees with the approved drawings.	18.5.5
23. Exit marking audible notification appliances	X	Semiannual	Verify location and condition.	

Table 14.3.1 ***Continued***

Component	Initial Acceptance	Periodic Frequency	Method	Reference
24. Reserved				
25. Area of refuge two-way communication system	X	Annual	Verify location and condition.	
26. Reserved				
27. Supervising station alarm systems — receivers				
(a) Signal receipt	X	Daily	Verify receipt of signal.	
(b) Receivers	X	Annual	Verify location and normal condition.	
28. Public emergency alarm reporting system transmission equipment			Verify location and condition.	
(a) Publicly accessible alarm box	X	Semiannual		
(b) Auxiliary box	X	Annual		
(c) Master box				
(1) Manual operation	X	Semiannual		
(2) Auxiliary operation	X	Annual		
29. Reserved				
30. Mass notification system				
(a) Monitored for integrity			Verify a system normal condition.	
(1) Control equipment				
(i) Fuses	X	Annual		
(ii) Interfaces	X	Annual		
(iii) Lamps/LED	X	Annual		
(iv) Primary (main) power supply	X	Annual		
(2) Secondary power batteries	X	Annual		
(3) Initiating devices	X	Annual		
(4) Notification appliances	X	Annual		
(b) Not monitored for integrity; installed prior to adoption of the 2010 edition			Verify a system normal condition.	
(1) Control equipment				
(i) Fuses	X	Semiannual		
(ii) Interfaces	X	Semiannual		
(iii) Lamps/LED	X	Semiannual		
(iv) Primary (main) power supply	X	Semiannual		
(2) Secondary power batteries	X	Semiannual		
(3) Initiating devices	X	Semiannual		
(4) Notification appliances	X	Semiannual		
(c) Antenna	X	Annual	Verify location and condition.	
(d) Transceivers	X	Annual	Verify location and condition.	

Note: N/A = not applicable, no minimum requirement established.

14.3.2 Devices or equipment that is inaccessible for safety considerations (e.g., continuous process operations, energized electrical equipment, radiation, and excessive height) shall be permitted to be inspected during scheduled shutdowns if approved by the authority havingjurisdiction.

14.3.3 Extended intervals shall not exceed 18 months.

14.3.4 Initial and reacceptance inspections shall be made to ensure compliance with approved design documents and to ensure installation in accordance with this Code and other required installation standards.

14.3.5 Periodic visual inspections in accordance with Table 14.3.1 shall be made to assure that there are no changes that affect equipment performance.

14.4 Testing.

14.4.1 Initial Acceptance Testing.

14.4.1.1 All new systems shall be inspected and tested in accordance with the requirements of Chapter 14.

14.4.1.2 The authority having jurisdiction shall be notified prior to the initial acceptance test.

14.4.2* Reacceptance Testing.

14.4.2.1 When an initiating device, notification appliance, or control relay is added, it shall be functionally tested.

14.4.2.2 When an initiating device, notification appliance, or control relay is deleted, another device, appliance, or control relay on the circuit shall be operated.

14.4.2.3 When modifications or repairs to control equipment hardware are made, the control equipment shall be tested in accordance with Table 14.4.3.2, items 2(a) and 2(d).

14.4.2.4 When changes are made to site-specific software, the following shall apply:

(1) All functions known to be affected by the change, or identified by a means that indicates changes, shall be 100 percent tested.
(2) In addition, 10 percent of initiating devices that are not directly affected by the change, up to a maximum of 50 devices, also shall be tested and correct system operation shall be verified.
(3) A revised record of completion in accordance with 7.5.6 shall be prepared to reflect these changes.

14.4.2.5 Changes to the system executive software shall require a 10 percent functional test of the system, including a test of at least one device on each input and output circuit to verify critical system functions such as notification appliances, control functions, and off-premises reporting.

14.4.3* Test Methods.

14.4.3.1* At the request of the authority having jurisdiction, the central station facility installation shall be inspected for complete information regarding the central station system, including specifications, wiring diagrams, and floor plans that have been submitted for approval prior to installation of equipment and wiring.

14.4.3.2* Systems and associated equipment shall be tested according to Table 14.4.3.2.

Table 14.4.3.2 Testing

Component	Initial Acceptance	Periodic Frequency	Method
1. All equipment	X		See Table 14.3.1.
2. Control equipment and transponder			
(a) Functions	X	Annually	Verify correct receipt of alarm, supervisory, and trouble signals (inputs); operation of evacuation signals and auxiliary functions (outputs); circuit supervision, including detection of open circuits and ground faults; and power supply supervision for detection of loss of ac power and disconnection of secondary batteries.
(b) Fuses	X	Annually	Verify rating and supervision.
(c) Interfaced equipment	X	Annually	Verify integrity of single or multiple circuits providing interface between two or more control units. Test interfaced equipment connections by operating or simulating operation of the equipment being supervised. Verify signals required to be transmitted at the control unit.
(d) Lamps and LEDs	X	Annually	Illuminate lamps and LEDs.
(e) Primary (main) power supply	X	Annually	Test under maximum load, including all alarm appliances requiring simultaneous operation. Test redundant power supplies separately.
3. Fire alarm control unit trouble signals			
(a) Audible and visual	X	Annually	Verify operation of control unit trouble signals. Verify ring-back feature for systems using a trouble-silencing switch that requires resetting.
(b) Disconnect switches	X	Annually	If control unit has disconnect or isolating switches, verify performance of intended function of each switch. Verify receipt of trouble signal when a supervised function is disconnected.
(c) Ground-fault monitoring circuit	X	Annually	If the system has a ground detection feature, verify the occurrence of ground-fault indication whenever any installation conductor is grounded.
(d) Transmission of signals to off-premises location	X	Annually	Actuate an initiating device and verify receipt of alarm signal at the off-premises location. Create a trouble condition and verify receipt of a trouble signal at the off-premises location. Actuate a supervisory device and verify receipt of a supervisory signal at the off-premises location. If a transmission carrier is capable of operation under a single- or multiple-fault condition, activate an initiating device during such fault condition and verify receipt of an alarm signal and a trouble signal at the off-premises location.

Table 14.4.3.2 ***Continued***

Component	Initial Acceptance	Periodic Frequency	Method
4. Supervising station alarm systems — transmission equipment			
(a) All equipment	X	Annually	[a]Test all system functions and features in accordance with the equipment manufacturer's published instructions for correct operation in conformance with the applicable sections of Chapter 26. Except for DACT, actuate initiating device and verify receipt of the correct initiating device signal at the supervising station within 90 seconds. Upon completion of the test, restore the system to its functional operating condition. If test jacks are used, conduct the first and last tests without the use of the test jack.
(b) Digital alarm communicator transmitter (DACT)	X	Annually	Except for DACTs installed prior to adoption of the 2013 edition of NFPA 72 that are connected to a telephone line (number) that is also supervised for adverse conditions by a derived local channel, ensure connection of the DACT to two separate means of transmission. Test DACT for line seizure capability by initiating a signal while using the telephone line (primary line for DACTs using two telephone lines) for a telephone call. Ensure that the call is interrupted and that the communicator connects to the digital alarm receiver. Verify receipt of the correct signal at the supervising station. Verify each transmission attempt is completed within 90 seconds from going off-hook to on-hook. Disconnect the telephone line (primary line for DACTs using two telephone lines) from the DACT. Verify indication of the DACT trouble signal occurs at the premises fire alarm control unit within 4 minutes of detection of the fault. Verify receipt of the telephone line trouble signal at the supervising station. Restore the telephone line (primary line for DACTs using two telephone lines), reset the fire alarm control unit, and verify that the telephone line fault trouble signal returns to normal. Verify that the supervising station receives the restoral signal from the DACT. Disconnect the secondary means of transmission from the DACT. Verify indication of the DACT trouble signal occurs at the premises fire alarm control unit within 4 minutes of detection of the fault. Verify receipt of the secondary means trouble signal at the supervising station. Restore the secondary means of transmission, reset the fire alarm control unit, and verify that the trouble signal returns to normal. Verify that the supervising station receives the restoral signal from the secondary transmitter. Cause the DACT to transmit a signal to the DACR while a fault in the telephone line (number) (primary line for DACTs using two telephone lines) is simulated. Verify utilization of the secondary communication path by the DACT to complete the transmission to the DACR.
(c) Digital alarm radio transmitter (DART)	X	Annually	Disconnect the primary telephone line. Verify transmission of a trouble signal to the supervising station by the DART occurs within 4 minutes.
(d) McCulloh transmitter	X	Annually	Actuate initiating device. Verify production of not less than three complete rounds of not less than three signal impulses each by the McCulloh transmitter. If end-to-end metallic continuity is present and with a balanced circuit, cause each of the following four transmission channel fault conditions in turn, and verify receipt of correct signals at the supervising station: (1) Open (2) Ground (3) Wire-to-wire short (4) Open and ground If end-to-end metallic continuity is not present and with a properly balanced circuit, cause each of the following three transmission channel fault conditions in turn, and verify receipt of correct signals at the supervising station: (1) Open (2) Ground (3) Wire-to-wire short

(continues)

Table 14.4.3.2 *Continued*

Component	Initial Acceptance	Periodic Frequency	Method
(e) Radio alarm transmitter (RAT)	X	Annually	Cause a fault between elements of the transmitting equipment. Verify indication of the fault at the protected premises, or transmission of trouble signal to the supervising station.
(f) Performance-based technologies	X	Annually	Perform tests to ensure the monitoring of integrity of the transmission technology and technology path. Where shared communications equipment is used as permitted by 26.6.3.1.14, provided secondary (standby) power sources shall be tested in accordance with Table 14.4.3.2, item 7, 8, or 9, as applicable. Where a single communications path is used, disconnect the communication path. Manually initiate an alarm signal transmission or allow the check-in (handshake) signal to be transmitted automatically.[b] Verify the premises unit annunciates the failure within 200 seconds of the transmission failure. Restore the communication path. Where multiple communication paths are used, disconnect both communication paths. Manually initiate an alarm signal transmission. Verify the premises control unit annunciates the failure within 200 seconds of the transmission failure. Restore both communication paths.
5. Emergency communications equipment			
(a) Amplifier/tone generators	X	Annually	Verify correct switching and operation of backup equipment.
(b) Call-in signal silence	X	Annually	Operate/function and verify receipt of correct visual and audible signals at control unit.
(c) Off-hook indicator (ring down)	X	Annually	Install phone set or remove phone from hook and verify receipt of signal at control unit.
(d) Phone jacks	X	Annually	Visually inspect phone jack and initiate communications path through jack.
(e) Phone set	X	Annually	Activate each phone set and verify correct operation.
(f) System performance	X	Annually	Operate the system with a minimum of any five handsets simultaneously. Verify voice quality and clarity.
6. Engine-driven generator	X	Monthly	If an engine-driven generator dedicated to the system is used as a required power source, verify operation of the generator in accordance with NFPA 110 by the building owner.
7. Secondary (standby) power supply[c]	X	Annually	Disconnect all primary (main) power supplies and verify the occurrence of required trouble indication for loss of primary power. Measure or verify the system's standby and alarm current demand and verify the ability of batteries to meet standby and alarm requirements using manufacturer's data. Operate general alarm systems a minimum of 5 minutes and emergency voice communications systems for a minimum of 15 minutes. Reconnect primary (main) power supply at end of test.
8. Uninterruptible power supply (UPS)	X	Annually	If a UPS system dedicated to the system is used as a required power source, verify by the building owner operation of the UPS system in accordance with NFPA 111.
9. Battery tests			Prior to conducting any battery testing, verify by the person conducting the test, that all system software stored in volatile memory is protected from loss.
(a) Lead-acid type			
(1) Battery replacement	X	Annually	Replace batteries in accordance with the recommendations of the alarm equipment manufacturer or when the recharged battery voltage or current falls below the manufacturer's recommendations.
(2) Charger test	X	Annually	With the batteries fully charged and connected to the charger, measure the voltage across the batteries with a voltmeter. Verify the voltage is 2.30 volts per cell ±0.02 volts at 77°F (25°C) or as specified by the equipment manufacturer.
(3) Discharge test	X	Annually	With the battery charger disconnected, load test the batteries following the manufacturer's recommendations. Verify the voltage level does not fall below the levels specified. Load testing can be by means of an artificial load equal to the full fire alarm load connected to the battery.
(4) Load voltage test	X	Semiannually	With the battery charger disconnected, load test the batteries following the manufacturer's recommendations. Verify the voltage level does not fall below the levels specified. Load testing can be by means of an artificial load equal to the full fire alarm load connected to the battery. Verify the battery does not fall below 2.05 volts per cell under load.

Table 14.4.3.2 ***Continued***

Component	Initial Acceptance	Periodic Frequency	Method
(5) Specific gravity	X	Semiannually	Measure as required the specific gravity of the liquid in the pilot cell or all of the cells. Verify the specific gravity is within the range specified by the manufacturer. Although the specified specific gravity varies from manufacturer to manufacturer, a range of 1.205–1.220 is typical for regular lead-acid batteries, while 1.240–1.260 is typical for high-performance batteries. Do not use a hydrometer that shows only a pass or fail condition of the battery and does not indicate the specific gravity, because such a reading does not give a true indication of the battery condition.
(b) Nickel-cadmium type			
(1) Battery replacement	X	Annually	Replace batteries in accordance with the recommendations of the alarm equipment manufacturer or when the recharged battery voltage or current falls below the manufacturer's recommendations.
(2) Charger test[d]	X	Annually	With the batteries fully charged and connected to the charger, place an ampere meter in series with the battery under charge. Verify the charging current is in accordance with the manufacturer's recommendations for the type of battery used. In the absence of specific information, use $\frac{1}{30}$ to $\frac{1}{25}$ of the battery rating.
(3) Discharge test	X	Annually	With the battery charger disconnected, load test the batteries following the manufacturer's recommendations. Verify the voltage level does not fall below the levels specified. Load testing can be by means of an artificial load equal to the full fire alarm load connected to the battery.
(4) Load voltage test	X	Semiannually	With the battery charger disconnected, load test the batteries following the manufacturer's recommendations. Verify the voltage level does not fall below the levels specified. Load testing can be by means of an artificial load equal to the full fire alarm load connected to the battery. Verify the float voltage for the entire battery is 1.42 volts per cell, nominal, under load. If possible, measure cells individually.
(c) Sealed lead-acid type			
(1) Battery replacement	X	Annually	Replace batteries in accordance with the recommendations of the alarm equipment manufacturer or when the recharged battery voltage or current falls below the manufacturer's recommendations.
(2) Charger test	X	Annually	With the batteries fully charged and connected to the charger, measure the voltage across the batteries with a voltmeter. Verify the voltage is 2.30 volts per cell ±0.02 volts at 77°F (25°C) or as specified by the equipment manufacturer.
(3) Discharge test	X	Annually	With the battery charger disconnected, load test the batteries following the manufacturer's recommendations. Verify the voltage level does not fall below the levels specified. Load testing can be by means of an artificial load equal to the full fire alarm load connected to the battery.
(4) Load voltage test	X	Semiannually	Verify the battery performs under load, in accordance with the battery manufacturer's specifications.
10. Public emergency alarm reporting system — wired system	X	Daily	Manual tests of the power supply for public reporting circuits shall be made and recorded at least once during each 24-hour period. Such tests shall include the following: (1) Current strength of each circuit. Changes in current of any circuit exceeding 10 percent shall be investigated immediately. (2) Voltage across terminals of each circuit inside of terminals of protective devices. Changes in voltage of any circuit exceeding 10 percent shall be investigated immediately. (3)[e] Voltage between ground and circuits. If this test shows a reading in excess of 50 percent of that shown in the test specified in (2), the trouble shall be immediately located and cleared. Readings in excess of 25 percent shall be given early attention. These readings shall be taken with a calibrated voltmeter of not more than 100 ohms resistance per volt. Systems in which each circuit is supplied by an independent current source (Forms 3 and 4) require tests between ground and each side of each circuit. Common current source systems (Form 2) require voltage tests between ground and each terminal of each battery and other current source.

(continues)

Table 14.4.3.2 *Continued*

Component	Initial Acceptance	Periodic Frequency	Method
			(4) Ground current reading shall be permitted in lieu of (3). If this method of testing is used, all grounds showing a current reading in excess of 5 percent of the supplied line current shall be given immediate attention. (5) Voltage across terminals of common battery on switchboard side of fuses. (6) Voltage between common battery terminals and ground. Abnormal ground readings shall be investigated immediately. Tests specified in (5) and (6) shall apply only to those systems using a common battery. If more than one common battery is used, each common battery shall be tested.
11. Remote annunciators	X	Annually	Verify the correct operation and identification of annunciators. If provided, verify the correct operation of annunciator under a fault condition.
12. Reserved			
13. Reserved			
14. Reserved			
15. Conductors — metallic			
(a) Stray voltage	X	N/A	Test all installation conductors with a volt/ohmmeter to verify that there are no stray (unwanted) voltages between installation conductors or between installation conductors and ground. Verify the maximum allowable stray voltage does not exceed 1 volt ac/dc, unless a different threshold is specified in the published manufacturer's instructions for the installed equipment.
(b) Ground faults	X	N/A	Test all installation conductors, other than those intentionally and permanently grounded, for isolation from ground per the installed equipment manufacturer's published instructions.
(c) Short-circuit faults	X	N/A	Test all installation conductors, other than those intentionally connected together, for conductor-to-conductor isolation per the published manufacturer's instructions for the installed equipment. Also test these same circuits conductor-to-ground.
(d) Loop resistance	X	N/A	With each initiating and indicating circuit installation conductor pair short-circuited at the far end, measure and record the resistance of each circuit. Verify that the loop resistance does not exceed the limits specified in the published manufacturer's instructions for the installed equipment.
(e) Circuit integrity	X	N/A	For initial and reacceptance testing, confirm the introduction of a fault in any circuit monitored for integrity results in a trouble indication at the fire alarm control unit. Open one connection at not less than 10 percent of the initiating devices, notification appliances and controlled devices on every initiating device circuit, notification appliance circuit, and signaling line circuit. Confirm all circuits perform as indicated in Sections 23.5, 23.6, and 23.7.
	N/A	Annually	For periodic testing, test each initiating device circuit, notification appliance circuit, and signaling line circuit for correct indication at the control unit. Confirm all circuits perform as indicated in Sections 23.5, 23.6, and 23.7.
16. Conductors — nonmetallic			
(a) Fiber optics	X	N/A	Test the fiber-optic transmission line by the use of an optical power meter or by an optical time domain reflectometer used to measure the relative power loss of the line. Test result data must meet or exceed ANSI/TIA 568-C.3, *Optical Fiber Cabling Components Standard*, related to fiber-optic lines and connection/splice losses and the control unit manufacturer's published specifications.
(b) Circuit integrity	X	N/A	For initial and reacceptance testing, confirm the introduction of a fault in any circuit monitored for integrity results in a trouble indication at the fire alarm control unit. Open one connection at not less than 10 percent of the initiating devices, notification appliances, and controlled devices on every initiating device circuit, notification appliance circuit, and signaling line circuit. Confirm all circuits perform as indicated in Sections 23.5, 23.6, and 23.7.
	N/A	Annually	For periodic testing, test each initiating device circuit, notification appliance circuit, and signaling line circuit for correct indication at the control unit. Confirm all circuits perform as indicated in Sections 23.5, 23.6, and 23.7.

Table 14.4.3.2 *Continued*

Component	Initial Acceptance	Periodic Frequency	Method
17. Initiating devices[f]			
(a) Electromechanical releasing device			
(1) Nonrestorable-type link	X	Annually	Verify correct operation by removal of the fusible link and operation of the associated device. Lubricate any moving parts as necessary.
(2) Restorable-type link[g]	X	Annually	Verify correct operation by removal of the fusible link and operation of the associated device. Lubricate any moving parts as necessary.
(b) Fire extinguishing system(s) or suppression system(s) alarm switch	X	Annually	Operate the switch mechanically or electrically and verify receipt of signal by the fire alarm control unit.
(c) Fire–gas and other detectors	X	Annually	Test fire–gas detectors and other fire detectors as prescribed by the manufacturer and as necessary for the application.
(d) Heat detectors			
(1) Fixed-temperature, rate-of-rise, rate of compensation, restorable line, spot type (excluding pneumatic tube type)	X	Annually (see 14.4.4.5)	Perform heat test with a listed and labeled heat source or in accordance with the manufacturer's published instructions. Assure that the test method for the installed equipment does not damage the nonrestorable fixed-temperature element of a combination rate-of-rise/fixed-temperature element detector.
(2) Fixed-temperature, nonrestorable line type	X	Annually	Do not perform heat test. Test functionality mechanically and electrically. Measure and record loop resistance. Investigate changes from acceptance test.
(3) Fixed-temperature, nonrestorable spot type	X	See Method	After 15 years from initial installation, replace all devices or have 2 detectors per 100 laboratory tested. Replace the 2 detectors with new devices. If a failure occurs on any of the detectors removed, remove and test additional detectors to determine either a general problem involving faulty detectors or a localized problem involving 1 or 2 defective detectors. If detectors are tested instead of replaced, repeat tests at intervals of 5 years.
(4) Nonrestorable (general)	X	Annually	Do not perform heat tests. Test functionality mechanically and electrically.
(5) Restorable line type, pneumatic tube only	X	Annually	Perform heat tests (where test chambers are in circuit), with a listed and labeled heat source or in accordance with the manufacturer's published instructions of the detector or conduct a test with pressure pump.
(6) Single- and multiple-station heat alarms	X	Annually	Conduct functional tests according to manufacturer's published instructions. Do not test nonrestorable heat detectors with heat.
(e) Manual fire alarm boxes	X	Annually	Operate manual fire alarm boxes per the manufacturer's published instructions. Test both key-operated presignal and general alarm manual fire alarm boxes.
(f) Radiant energy fire detectors	X	Semiannually	Test flame detectors and spark/ember detectors in accordance with the manufacturer's published instructions to determine that each detector is operative. Determine flame detector and spark/ember detector sensitivity using any of the following: (1) Calibrated test method (2) Manufacturer's calibrated sensitivity test instrument (3) Listed control unit arranged for the purpose (4) Other approved calibrated sensitivity test method that is directly proportional to the input signal from a fire, consistent with the detector listing or approval If designed to be field adjustable, replace detectors found to be outside of the approved range of sensitivity or adjust to bring them into the approved range. Do not determine flame detector and spark/ember detector sensitivity using a light source that administers an unmeasured quantity of radiation at an undefined distance from the detector.
(g) Smoke detectors — functional test			
(1) In other than one- and two-family dwellings, system detectors	X	Annually	[h]Test smoke detectors in place to ensure smoke entry into the sensing chamber and an alarm response. Use smoke or a listed and labeled product acceptable to the manufacturer or in accordance with their published instructions. Other methods listed in the manufacturer's published instructions that ensure smoke entry from the protected area, through the vents, into the sensing chamber can be used.

(continues)

Table 14.4.3.2 ***Continued***

Component	Initial Acceptance	Periodic Frequency	Method
(2) Single- and multiple-station smoke alarms connected to protected premises systems	X	Annually	Perform a functional test on all single- and multiple-station smoke alarms connected to a protected premises fire alarm system by putting the smoke alarm into an alarm condition and verifying that the protected premises system receives a supervisory signal and does not cause a fire alarm signal.
(3) System smoke detectors used in one- and two-family dwellings	X	Annually	Conduct functional tests according to manufacturer's published instructions.
(4) Air sampling	X	Annually	Test with smoke or a listed and labeled product acceptable to the manufacturer or in accordance with their published instructions. Test from the end sampling port or point on each pipe run. Verify airflow through all other ports or points.
(5) Duct type	X	Annually	In addition to the testing required in Table 14.4.3.2(g)(1) and Table 14.4.3.2(h), test duct smoke detectors that use sampling tubes to ensure that they will properly sample the airstream in the duct using a method acceptable to the manufacturer or in accordance with their published instructions.
(6) Projected beam type	X	Annually	Test the detector by introducing smoke, other aerosol, or an optical filter into the beam path.
(7) Smoke detector with built-in thermal element	X	Annually	Operate both portions of the detector independently as described for the respective devices.
(8) Smoke detectors with control output functions	X	Annually	Verify that the control capability remains operable even if all of the initiating devices connected to the same initiating device circuit or signaling line circuit are in an alarm state.
(h) Smoke detectors — sensitivity testing			
In other than one- and two-family dwellings, system detectors	N/A	See 14.4.4.3	[i]Perform any of the following tests to ensure that each smoke detector is within its listed and marked sensitivity range: (1) Calibrated test method (2) Manufacturer's calibrated sensitivity test instrument (3) Listed control equipment arranged for the purpose (4) Smoke detector/control unit arrangement whereby the detector causes a signal at the control unit when its sensitivity is outside its listed sensitivity range (5) Other calibrated sensitivity test method approved by the authority having jurisdiction
(i) Carbon monoxide detectors/carbon monoxide alarms for the purposes of fire detection	X	Annually	Test the devices in place to ensure CO entry to the sensing chamber by introduction through the vents, to the sensing chamber of listed and labeled product acceptable to the manufacturer or in accordance with their published instructions.
(j) Initiating devices, supervisory			
(1) Control valve switch	X	Semiannual	Operate valve and verify signal receipt to be within the first two revolutions of the handwheel or within one-fifth of the travel distance, or per the manufacturer's published instructions.
(2) High- or low-air pressure switch	X	Annually	Operate switch and verify receipt of signal is obtained where the required pressure is increased or decreased a maximum 10 psi (70 kPa) from the required pressure level.
(3) Room temperature switch	X	Annually	Operate switch and verify receipt of signal to indicate the decrease in room temperature to 40°F (4.4°C) and its restoration to above 40°F (4.4°C).
(4) Water level switch	X	Annually	Operate switch and verify receipt of signal indicating the water level raised or lowered a maximum 3 in. (70 mm) from the required level within a pressure tank, or a maximum 12 in. (300 mm) from the required level of a nonpressure tank. Also verify its restoral to required level.
(5) Water temperature switch	X	Annually	Operate switch and verify receipt of signal to indicate the decrease in water temperature to 40°F (4.4°C) and its restoration to above 40°F (4.4°C).
(k) Mechanical, electrosonic, or pressure-type waterflow device	X	Semiannually	Water shall be flowed through an inspector's test connection indicating the flow of water equal to that from a single sprinkler of the smallest orifice size installed in the system for wet-pipe systems, or an alarm test bypass connection for dry-pipe, pre-action, or deluge systems in accordance with NFPA 25.

Table 14.4.3.2 ***Continued***

Component	Initial Acceptance	Periodic Frequency	Method
(l) Multi-sensor fire detector or multi-criteria fire detector or combination fire detector	X	Annually	Test each of the detection principles present within the detector (e.g., smoke/heat/CO, etc.) independently for the specific detection principle, regardless of the configuration status at the time of testing. Also test each detector in accordance with the published manufacturer's instructions. Test individual sensors together if the technology allows individual sensor responses to be verified. Perform tests as described for the respective devices by introduction of the physical phenomena to the sensing chamber of element. An electronic check (magnets, analog values, etc.) is not sufficient to comply with this requirement. Verify by using the detector manufacturer's published instructions that the test gas used will not impair the operation of either sensing chamber of a multisensor, multicriteria, or combination fire detector. Confirm the result of each sensor test through indication at the detector or control unit. Where individual sensors cannot be tested individually, test the primary sensor.[j] Record all tests and results.
18. Special hazard equipment			
(a) Abort switch (dead-man type)	X	Annually	Operate abort switch and verify correct sequence and operation.
(b) Abort switch (recycle type)	X	Annually	Operate abort switch and verify development of correct matrix with each sensor operated.
(c) Abort switch (special type)	X	Annually	Operate abort switch and verify correct sequence and operation in accordance with authority having jurisdiction. Observe sequencing as specified on as-built drawings or in system owner's manual.
(d) Cross-zone detection circuit	X	Annually	Operate one sensor or detector on each zone. Verify occurrence of correct sequence with operation of first zone and then with operation of second zone.
(e) Matrix-type circuit	X	Annually	Operate all sensors in system. Verify development of correct matrix with each sensor operated.
(f) Release solenoid circuit[k]	X	Annually	Verify operation of solenoid.
(g) Squibb release circuit	X	Annually	Use AGI flashbulb or other test light approved by the manufacturer. Verify operation of flashbulb or light.
(h) Verified, sequential, or counting zone circuit	X	Annually	Operate required sensors at a minimum of four locations in circuit. Verify correct sequence with both the first and second detector in alarm.
(i) All above devices or circuits or combinations thereof	X	Annually	Verify supervision of circuits by creating an open circuit.
19. Combination systems			
(a) Fire extinguisher electronic monitoring device/system	X	Annually	Test communication between the device connecting the fire extinguisher electronic monitoring device/system and the fire alarm control unit to ensure proper signals are received at the fire alarm control unit and remote annunciator(s) if applicable.
(b) Carbon monoxide[l] device/system	X	Annually	Test communication between the device connecting the carbon monoxide device/system and the fire alarm control unit to ensure proper signals are received at the fire alarm control unit and remote annunciator(s) if applicable.
20. Interface equipment[m]	X	See 14.4.4.4	Test interface equipment connections by operating or simulating the equipment being supervised. Verify signals required to be transmitted are received at the control unit. Test frequency for interface equipment is the same as the frequency required by the applicable NFPA standard(s) for the equipment being supervised.
21. Guard's tour equipment	X	Annually	Test the device in accordance with the manufacturer's published instructions.
22. Alarm notification appliances			
(a) Audible[n]	X	N/A	For initial and reacceptance testing, measure sound pressure levels for signals with a sound level meter meeting ANSI S1.4a, *Specifications for Sound Level Meters,* Type 2 requirements. Measure sound pressure levels throughout the protected area to confirm that they are in compliance with Chapter 18. Set the sound level meter in accordance with ANSI S3.41, *American National Standard Audible Evacuation Signal,* using the time-weighted characteristic F (FAST).

(continues)

Table 14.4.3.2 ***Continued***

Component	Initial Acceptance	Periodic Frequency	Method
	N/A	Annually	°For periodic testing, verify the operation of the notification appliances.
(b) Audible textual notification appliances (speakers and other appliances to convey voice messages)	X	N/A	For initial and reacceptance testing, measure sound pressure levels for signals with a sound level meter meeting ANSI S1.4a, *Specifications for Sound Level Meters,* Type 2 requirements. Measure sound pressure levels throughout the protected area to confirm that they are in compliance with Chapter 18. Set the sound level meter in accordance with ANSI S3.41, *American National Standard Audible Evacuation Signal,* using the time-weighted characteristic F (FAST). Verify audible information to be distinguishable and understandable and in compliance with 14.4.11.
	N/A	Annually	°For periodic testing, verify the operation of the notification appliances.
(c) Visible	X	N/A	Perform initial and reacceptance testing in accordance with the manufacturer's published instructions. Verify appliance locations to be per approved layout and confirm that no floor plan changes affect the approved layout. Verify that the candela rating marking agrees with the approved drawing. Confirm that each appliance flashes.
	N/A	Annually	For periodic testing, verify that each appliance flashes.
23. Exit marking audible notification appliance	X	Annually	Perform tests in accordance with manufacturer's published instructions.
24. Emergency control functions[p]	X	Annually	For initial, reacceptance, and periodic testing, verify emergency control function interface device activation. Where an emergency control function interface device is disabled or disconnected during initiating device testing, verify that the disabled or disconnected emergency control function interface device has been properly restored.
25. Area of refuge two-way communication system	X	Annually	Use the manufacturer's published instructions and the as-built drawings provided by the system supplier to verify correct operation after the initial testing phase has been performed by the supplier or by the supplier's designated representative. Test the two-way communication system to verify operation and receipt of visual and audible signals at the transmitting unit and the receiving unit, respectively. Operate systems with more than five stations with a minimum of five stations operating simultaneously. Verify voice quality and clarity. Verify directions for the use of the two-way communication system, instructions for summoning assistance via the two-way communication system, and written identification of the location is posted adjacent to the two-way communication system. Verify that all remote stations are readily accessible. Verify the timed automatic communications capability to connect with a constantly attended monitoring location per 24.5.3.4.
26. Special procedures			
(a) Alarm verification	X	Annually	Verify time delay and alarm response for smoke detector circuits identified as having alarm verification.
(b) Multiplex systems	X	Annually	Verify communications between sending and receiving units under both primary and secondary power. Verify communications between sending and receiving units under open-circuit and short-circuit trouble conditions. Verify communications between sending and receiving units in all directions where multiple communications pathways are provided. If redundant central control equipment is provided, verify switchover and all required functions and operations of secondary control equipment. Verify all system functions and features in accordance with manufacturer's published instructions.
27. Supervising station alarm systems — receiving equipment			
(a) All equipment	X	Monthly	Perform tests on all system functions and features in accordance with the equipment manufacturer's published instructions for correct operation in conformance with the applicable sections of Chapter 26. Actuate initiating device and verify receipt of the correct initiating device signal at the supervising station within 90 seconds. Upon completion of the test, restore the system to its functional operating condition.

Table 14.4.3.2 ***Continued***

Component	Initial Acceptance	Periodic Frequency	Method
			If test jacks are used, perform the first and last tests without the use of the test jack.
(b) Digital alarm communicator receiver (DACR)	X	Monthly	Disconnect each transmission means in turn from the DACR, and verify audible and visual annunciation of a trouble signal in the supervising station. Cause a signal to be transmitted on each individual incoming DACR line (path) at least once every 6 hours (24 hours for DACTs installed prior to adoption of the 2013 edition of *NFPA 72*). Verify receipt of these signals.
(c) Digital alarm radio receiver (DARR)	X	Monthly	Cause the following conditions of all DARRs on all subsidiary and repeater station receiving equipment. Verify receipt at the supervising station of correct signals for each of the following conditions: (1) AC power failure of the radio equipment (2) Receiver malfunction (3) Antenna and interconnecting cable failure (4) Indication of automatic switchover of the DARR (5) Data transmission line failure between the DARR and the supervising or subsidiary station
(d) McCulloh systems	X	Monthly	Test and record the current on each circuit at each supervising and subsidiary station under the following conditions: (1) During functional operation (2) On each side of the circuit with the receiving equipment conditioned for an open circuit Cause a single break or ground condition on each transmission channel. If such a fault prevents the functioning of the circuit, verify receipt of a trouble signal. Cause each of the following conditions at each of the supervising or subsidiary stations and all repeater station radio transmitting and receiving equipment; verify receipt of correct signals at the supervising station: (1) RF transmitter in use (radiating) (2) AC power failure supplying the radio equipment (3) RF receiver malfunction (4) Indication of automatic switchover
(e) Radio alarm supervising station receiver (RASSR) and radio alarm repeater station receiver (RARSR)	X	Monthly	Cause each of the following conditions at each of the supervising or subsidiary stations and all repeater station radio transmitting and receiving equipment; verify receipt of correct signals at the supervising station: (1) AC power failure supplying the radio equipment (2) RF receiver malfunction (3) Indication of automatic switchover, if applicable
(f) Private microwave radio systems	X	Monthly	Cause each of the following conditions at each of the supervising or subsidiary stations and all repeater station radio transmitting and receiving equipment; verify receipt of correct signals at the supervising station: (1) RF transmitter in use (radiating) (2) AC power failure supplying the radio equipment (3) RF receiver malfunction (4) Indication of automatic switchover
(g) Performance-based technologies	X	Monthly	Perform tests to ensure the monitoring of integrity of the transmission technology and technology path. Where a single communications path is used, disconnect the communication path. Verify that failure of the path is annunciated at the supervising station within 60 minutes of the failure (within 5 minutes for communication equipment installed prior to adoption of the 2013 edition of *NFPA 72*). Restore the communication path. Where multiple communication paths are used, disconnect both communication paths and confirm that failure of the path is annunciated at the supervising station within not more than 6 hours of the failure (within 24 hours for communication equipment installed prior to adoption of the 2013 edition of *NFPA 72*). Restore both communication paths.

(continues)

Table 14.4.3.2 ***Continued***

Component	Initial Acceptance	Periodic Frequency	Method
28. Public emergency alarm reporting system transmission equipment			
(a) Publicly accessible alarm box	X	Semiannually	Actuate publicly accessible initiating device(s) and verify receipt of not less than three complete rounds of signal impulses. Perform this test under normal circuit conditions. If the device is equipped for open circuit operation (ground return), test it in this condition as one of the semiannual tests.
(b) Auxiliary box	X	Annually	Test each initiating circuit of the auxiliary box by actuation of a protected premises initiating device connected to that circuit. Verify receipt of not less than three complete rounds of signal impulses.
(c) Master box			
(1) Manual operation	X	Semiannually	Perform the tests prescribed for 28(a).
(2) Auxiliary operation	X	Annually	Perform the tests prescribed for 28(b).
29. Low-power radio (wireless systems)	X	N/A	The following procedures describe additional acceptance and reacceptance test methods to verify wireless protection system operation: (1) Use the manufacturer's published instructions and the as-built drawings provided by the system supplier to verify correct operation after the initial testing phase has been performed by the supplier or by the supplier's designated representative. (2) Starting from the functional operating condition, initialize the system in accordance with the manufacturer's published instructions. Confirm the alternative communications path exists between the wireless control unit and peripheral devices used to establish initiation, indication, control, and annunciation. Test the system for both alarm and trouble conditions. (3) Check batteries for all components in the system monthly unless the control unit checks all batteries and all components daily.
30. Mass notification systems			
(a) Functions	X	Annually	At a minimum, test control equipment to verify correct receipt of alarm, supervisory, and trouble signals (inputs); operation of evacuation signals and auxiliary functions (outputs); circuit supervision, including detection of open circuits and ground faults; and power supply supervision for detection of loss of ac power and disconnection of secondary batteries.
(b) Fuses	X	Annually	Verify the rating and supervision.
(c) Interfaced equipment	X	Annually	Verify integrity of single or multiple circuits providing interface between two or more control units. Test interfaced equipment connections by operating or simulating operation of the equipment being supervised. Verify signals required to be transmitted at the control unit.
(d) Lamps and LEDs	X	Annually	Illuminate lamps and LEDs.
(e) Primary (main) power supply	X	Annually	Disconnect all secondary (standby) power and test under maximum load, including all alarm appliances requiring simultaneous operation. Reconnect all secondary (standby) power at end of test. For redundant power supplies, test each separately.
(f) Audible textual notification appliances (speakers and other appliances to convey voice messages)	X	Annually	Measure sound pressure level with a sound level meter meeting ANSI S1.4a, *Specifications for Sound Level Meters,* Type 2 requirements. Measure and record levels throughout protected area. Set the sound level meter in accordance with ANSI S3.41, *American National Standard Audible Evacuation Signal,* using the time-weighted characteristic F (FAST). Record the maximum output when the audible emergency evacuation signal is on. Verify audible information to be distinguishable and understandable.
(g) Visible	X	Annually	Perform test in accordance with manufacturer's published instructions. Verify appliance locations to be per approved layout and confirm that no floor plan changes affect the approved layout. Verify that the candela rating marking agrees with the approved drawing. Confirm that each appliance flashes.
(h) Control unit functions and no diagnostic failures are indicated	X	Annually	Review event log file and verify that the correct events were logged. Review system diagnostic log file; correct deficiencies noted in file. Delete unneeded log files. Delete unneeded error files. Verify that sufficient free disk space is available. Verify unobstructed flow of cooling air is available. Change/clean filters, cooling fans, and intake vents.
(i) Control unit reset	X	Annually	Power down the central control unit computer and restart it.
(j) Control unit security	X	Annually	If remote control software is loaded onto the system, verify that it is disabled to prevent unauthorized system access.
(k) Audible/visible functional test	X	Annually	Send out an alert to a diverse set of predesignated receiving devices and confirm receipt. Include at least one of each type of receiving device.

Table 14.4.3.2 ***Continued***

Component	Initial Acceptance	Periodic Frequency	Method
(l) Software backup	X	Annually	Make full system software backup. Rotate backups based on accepted practice at site.
(m) Secondary power test	X	Annually	Disconnect ac power. Verify the ac power failure alarm status on central control equipment. With ac power disconnected, verify battery voltage under load.
(n) Wireless signals	X	Annually	Check forward/reflected radio power is within specifications.
(o) Antenna	X	Annually	Check forward/reflected radio power is within specifications. Verify solid electrical connections with no observable corrosion.
(p) Transceivers	X	Annually	Verify proper operation and mounting is not compromised.

[a]Some transmission equipment (such as but not limited to cable modems, fiber-optic interface nodes, and VoIP interfaces) are typically powered by the building's electrical system using a secondary (standby) power supply that does not meet the requirements of this Code. This is intended to ensure that the testing authority verifies full secondary (standby) power as required by Chapter 10. Additionally, refer to Table 14.4.3.2, items 7 through 9, for secondary (standby) power supply testing.

[b]The automatic transmission of the check-in (handshake) signal can take up to 60 minutes to occur.

[c]See Table 14.4.3.2, Item 4(a) for the testing of transmission equipment.

[d]Example: 4000 mAh × 1⁄25 = 160 mA charging current at 77°F (25°C).

[e]The voltmeter sensitivity has been changed from 1000 ohms per volt to 100 ohms per volt so that the false ground readings (caused by induced voltages) are minimized.

[f]Initiating devices such as smoke detectors used for elevator recall, closing dampers, or releasing doors held in the open position that are permitted by the Code (*see* NFPA *101 9.6.3*) to initiate supervisory signals at the fire alarm control unit (FACU) should be tested at the same frequency (annual) as those devices when they are generating an alarm signal. They are not supervisory devices, but they initiate a supervisory signal at the FACU.

[g]Fusible thermal link detectors are commonly used to close fire doors and fire dampers. They are actuated by the presence of external heat, which causes a solder element in the link to fuse, or by an electric thermal device, which, when energized, generates heat within the body of the link, causing the link to fuse and separate.

[h]Note, it is customary for the manufacturer of the smoke detector to test a particular product from an aerosol provider to determine acceptability for use in smoke entry testing of their smoke detector/ smoke alarm. Magnets are not acceptable for smoke entry tests.

[i] There are some detectors that use magnets as a manufacturer's calibrated sensitivity test instrument.

[j]For example, it might not be possible to individually test the heat sensor in a thermally enhanced smoke detector.

[k]Manufacturer's instructions should be consulted to ensure a proper operational test. No suppression gas or agent is expected to be discharged during the test of the solenoid. See Test Plan of 14.2.10.

[l]Testing of CO device should be done to the requirements of NFPA 720.

[m]A monitor module installed on an interface device is not considered a supervisory device and therefore not subject to the quarterly testing frequency requirement. Test frequencies for interface devices should be in accordance with the applicable standard. For example, fire pump controller alarms such as phase reversal are required to be tested annually. If a monitor module is installed to identify phase reversal on the fire alarm control panel, it is not necessary to test for phase reversal four times a year.

[n]Chapter 18 would require 15 dB over average ambient sound for public mode spaces. Sometimes the ambient sound levels are different from what the design was based upon. Private operating mode would require 10 dB over average ambient at the location of the device.

[o]Where building, system, or occupancy changes have been observed, the owner should be notified of the changes. New devices might need to be installed and tested per the initial acceptance testing criteria.

[p]See A.14.4.3.2, and Table 14.4.3.2, Item 24.

14.4.3.3 Video image smoke and flame detectors shall be inspected, tested, and maintained in accordance with the manufacturer's published instructions.

14.4.3.4 Gas detectors shall be inspected, tested, and maintained in accordance with manufacturers' published instructions.

14.4.4* Testing Frequency. Unless otherwise permitted by other sections of this Code, testing shall be performed in accordance with the schedules in Table 14.4.3.2 or more often if required by the authority having jurisdiction.

14.4.4.1 Devices or equipment that are inaccessible for safety considerations (e.g., continuous process operations, energized electrical equipment, radiation, and excessive height) shall be permitted to be tested during scheduled shutdowns if approved by the authority having jurisdiction. Extended intervals shall not exceed 18 months.

14.4.4.2 If automatic testing is performed at least weekly by a remotely monitored fire alarm control unit specifically listed for the application, the manual testing frequency shall be permitted to be extended to annually. Table 14.4.3.2 shall apply.

14.4.4.3* In other than one- and two-family dwellings, sensitivity of smoke detectors shall be tested in accordance with 14.4.4.3.1 through 14.4.4.3.6.

14.4.4.3.1 Sensitivity shall be checked within 1 year after installation.

14.4.4.3.2 Sensitivity shall be checked every alternate year thereafter unless otherwise permitted by compliance with 14.4.4.3.3.

14.4.4.3.3 After the second required calibration test, if sensitivity tests indicate that the device has remained within its listed and marked sensitivity range (or 4 percent obscuration light gray smoke, if not marked), the length of time between calibration tests shall be permitted to be extended to a maximum of 5 years.

14.4.4.3.3.1 If the frequency is extended, records of nuisance alarms and subsequent trends of these alarms shall be maintained.

14.4.4.3.3.2 In zones or in areas where nuisance alarms show any increase over the previous year, calibration tests shall be performed.

14.4.4.3.4 Unless otherwise permitted by 14.4.4.3.5, smoke detectors found to have a sensitivity outside the listed and marked sensitivity range shall be cleaned and recalibrated or be replaced.

14.4.4.3.5 Smoke detectors listed as field adjustable shall be permitted to either be adjusted within the listed and marked sensitivity range, cleaned, and recalibrated, or be replaced.

14.4.4.3.6 The detector sensitivity shall not be tested or measured using any device that administers an unmeasured concentration of smoke or other aerosol into the detector or smoke alarm.

14.4.4.4 Test frequency of interfaced equipment shall be the same as specified by the applicable NFPA standards for the equipment being supervised.

14.4.4.5 Restorable fixed-temperature, spot-type heat detectors shall be tested in accordance with 14.4.4.5.1 through 14.4.4.5.4.

14.4.4.5.1 Two or more detectors shall be tested on each initiating circuit annually.

14.4.4.5.2 Different detectors shall be tested each year.

14.4.4.5.3 Test records shall be kept by the building owner specifying which detectors have been tested.

14.4.4.5.4 Within 5 years, each detector shall have been tested.

14.4.4.6* Circuit and pathway testing of each monitored circuit or pathway shall be conducted with initial acceptance or reacceptance testing to verify signals are indicated at the control unit for each of the abnormal conditions specified in Sections 23.5 through 23.7.

14.4.5 Single- and Multiple-Station Smoke Alarms. (SIG-HOU)

14.4.5.1 Single- and multiple-station smoke alarms and connected appliances shall be inspected, tested, and maintained in accordance with 14.4.5 and the manufacturer's published instructions.

14.4.5.2 Smoke alarms and connected appliances shall be inspected and tested at least monthly.

14.4.5.3* The responsibility for inspection, testing, and maintenance of smoke alarms and connected appliances shall be in accordance with 14.2.3.

14.4.5.4 Smoke alarms shall be replaced when they fail to respond to operability tests.

14.4.5.4.1 Smoke alarms shall not remain in service longer than 10 years from the date of manufacture, unless otherwise provided by the manufacturer's published instructions.

14.4.5.5* Combination smoke/carbon monoxide alarms shall be replaced when the end-of-life signal activates or 10 years from the date of manufacture, whichever comes first, unless otherwise provided by the manufacturer's published instructions.

14.4.5.6 Where batteries are used as a source of energy for smoke alarms or combination smoke/carbon monoxide alarms or single- and multiple-station smoke alarms, the batteries shall be replaced in accordance with the alarm equipment manufacturer's published instructions.

14.4.6 Household Fire Alarm Systems.

14.4.6.1 Testing. Household fire alarm systems shall be tested by a qualified service technician at least annually according to the methods of Table 14.4.3.2. The installing contractor shall be required to provide this information in writing to the customer upon completion of the system installation. To the extent that the fire alarm system is monitored offsite, the supervising station contractor shall provide notice of this requirement to the customer on a yearly basis. (SIG-HOU)

14.4.6.2 Maintenance. Maintenance of household fire alarm systems shall be conducted according to the manufacturer's published instructions. (SIG-HOU)

14.4.7 Circuits from Central Station. Circuits extending from the central station that have had no signal activity in the preceding 24 hours shall be tested at intervals of not more than 24 hours.

14.4.8 Public Emergency Alarm Reporting Systems.

14.4.8.1 Emergency power sources other than batteries shall be tested at least weekly in accordance with 14.4.8.1.1 and 14.4.8.1.2.